Critiquing Communication Innovation

US–CHINA RELATIONS IN THE AGE OF GLOBALIZATION

This series publishes the best, cutting-edge work tackling the opportunities and dilemmas of relations between the United States and China in the age of globalization. Books published in the series encompass both historical studies and contemporary analyses, and include both single-authored monographs and edited collections. Our books are comparative, offering in-depth communication-based analyses of how United States and Chinese officials, scholars, artists, and activists configure each other, portray the relations between the two nations, and depict their shared and competing interests. They are interdisciplinary, featuring scholarship that works in and across communication studies, rhetoric, literary criticism, film studies, cultural studies, international studies, and more. And they are international, situating their analyses at the crossroads of international communication and the nuances, complications, and opportunities of globalization as it has unfolded since World War II.

SERIES EDITOR
Stephen J. Hartnett, *University of Colorado Denver*

EDITORIAL BOARD
Rya Butterfield, *Nicholls State University*
Hsin-I Cheng, *Santa Clara University*
Patrick Shaou-Whea Dodge, *International College of Beijing*
Qingwen Dong, *University of the Pacific*
Mohan Dutta, *Massey University, New Zealand*
John Erni, *Hong Kong Baptist University*
Xiaohong Gao, *Communication University of China*
G. Thomas Goodnight, *University of Southern California*
Robert Hariman, *Northwestern University*
Rolien Hoyng, *Chinese University of Hong Kong*
Dongjing Kang, *Florida Gulf Coast University*
Lisa Keränen, *University of Colorado Denver*
Zhi Li, *Communication University of China*
Jingfang Liu, *Fudan University*
Xing Lu, *DePaul University*
Trevor Parry-Giles, *National Communication Association*
Phaedra C. Pezzullo, *University of Colorado Boulder*
Todd Sandel, *University of Macau*
Zhiwei Wang, *University of Zhengzhou*
Guobin Yang, *Annenberg School, University of Pennsylvania*
Yufang Zhang, *University of Shanghai*

Critiquing Communication Innovation

NEW MEDIA IN A MULTIPOLAR WORLD

Edited by Rolien Hoyng and Gladys Pak Lei Chong

MICHIGAN STATE UNIVERSITY PRESS | *East Lansing*

Copyright © 2022 by Michigan State University

∞ The paper used in this publication meets the minimum requirements of
ANSI/NISO Z39.48-1992 (R 1997) (Permanence of Paper).

Michigan State University Press
East Lansing, Michigan 48823-5245

Library of Congress Cataloging-in-Publication Data

Names: Hoyng, Rolien, editor. | Chong, Gladys Pak Lei, 1977– editor.
Title: Critiquing communication innovation : new media in a multipolar world /
  edited by Rolien Hoyng and Gladys Pak Lei Chong.
Description: East Lansing : Michigan State University Press, [2022]
  Series: US-China relations in the age of globalization
  Includes bibliographical references.
Identifiers: LCCN 2021039969 | ISBN 978-1-61186-429-8 (paperback)
  | ISBN 978-1-60917-698-3 (pdf) | ISBN 978-1-62895-466-1 (epub) | ISBN 978-1-62896-460-8 (Kindle)
Subjects: LCSH: Digital media—Social aspects. | Digital media—Technological innovations—China.
  | Telecommunication—Technological innovations—China—Influence.
Classification: LCC HM851 .C746 2022 | DDC 302.23/1—dc23/eng/20211013
LC record available at https://lccn.loc.gov/2021039969

Cover design by Erin Kirk
Cover image: Mobile phone repair shop in Hong Kong. Photo by Rolien Hoyng.

Visit Michigan State University Press at *www.msupress.org*

Contents

- vii   ACKNOWLEDGMENTS
- ix   INTRODUCTION: New, Old, and Uncertain Futures

- 1   Analyzing Chinese Platform Power: Infrastructure, Finance, and Geopolitics, *Lianrui Jia and David Nieborg*
- 29   Neoliberal Business-as-Usual or Post-Surveillance Capitalism with European Characteristics? The EU's General Data Protection Regulation in a Multipolar Internet, *Angela Daly*
- 55   The Global versus the National: Creativity in Turkey's Game Industry, *Serra Sezgin and Mutlu Binark*
- 81   *Making, New Shanzhai,* and Countercultural Values: Ethnographies of Contemporary, Innovative, and Entrepreneurial Digital Fabrication Communities in Shenzhen, China, *Daniel H. Mutibwa and Bingqing Xia*
- 115   Platformization of the Unlikely Creative Class: Kuaishou and Chinese Digital Cultural Production, *Jian Lin and Jeroen de Kloet*
- 141   Technology Translations between China and Ghana: The Case of Low-End Phone Design, *Miao Lu*
- 167   The Necropolitics of Innovation: Sensing Death in the Mediterranean Sea, *Monika Halkort*
- 187   CONCLUSION: Futures in the Plural, *Jack Linchuan Qiu*

- 197   CONTRIBUTORS

ON THE INTERSECTION OF EDGE BALL AND COURTESY:
NOTES ON SCHOLARSHIP IN THE AGE OF GLOBALIZATION

Like America or France or Brazil, China is a nation-state riven with fault-lines along region and race, ethnicity and education, linguistics and libido, gender and more general divisions. The US media tends to portray Chinese society as monolithic—billions of citizens censored into silence, its activists and dissidents fearful of retribution. The "reeducation" camps in Xinjiang, the "black prisons" that dot the landscape, and the Great Firewall prove this belief partially true. At the same time, there are more dissidents on the Chinese web than there are living Americans, and rallies, marches, strikes, and protests unfold in China each week. The nation is seething with action, much of it politically radical. What makes this political action so complicated and so difficult to comprehend is that no one knows how the state will respond on any given day. In his magnificent *Age of Ambition*, Evan Osnos notes that "Divining how far any individual [can] go in Chinese creative life [is] akin to carving a line in the sand at low tide in the dark." His tide metaphor is telling, for throughout Chinese history waves of what Deng Xiaoping called "openness and reform" have given way to repression, which can then swing back to what Chairman Mao once called "letting a hundred flowers bloom"—China thus offers a perpetually changing landscape, in which nothing is certain. For this reason, our Chinese colleagues and collaborators are taking great risks by participating in this book series. Authors in the "west" fear their books and articles will fail to find an audience; authors in China live in fear of a midnight knock at the door.

This series therefore strives to practice what Qingwen Dong calls "edge ball": Getting as close as possible to the boundary of what is sayable without crossing the line into being offensive. The image is borrowed from table tennis and depicts a shot that barely touches the line before ricocheting off the table; it counts as a point and is within the rules, yet the trajectory of the ball makes it almost impossible to hit a return shot. In the realm of scholarship and politics, playing "edge ball" means speaking truth to power while not provoking arrest—this is a murky game full of gray zones, allusions, puns, and sly references. What this means for our series is clear: Our authors do not censor themselves, but they do speak respectfully and cordially, showcasing research-based perspectives from their standpoints and their worldviews, thereby putting multiple vantage points into conversation. As our authors practice "edge ball," we hope our readers will savor these books with a similar sense of sophisticated and international generosity.

—Stephen J. Hartnett

Acknowledgments

This book was edited in Hong Kong, a city where the future has a difficult standing. The temporality of innovation used to provide a steady and predictable drumbeat to everyday life with the introduction of new gadgets, new storefronts, more smart-city buzz, more apps, and more science parks. This drumbeat maintained a promise that Hong Kong was heading toward a "modern" future, born from technology and innovation, that seemed more or less known. But Mainland China's transformation from so-called "copycat" to world-leading innovator is one development that has shaken up this banal sense of the familiar future. Amid the unsettling of the future, in Hong Kong but also elsewhere, this book on communication innovation and newness came into being.

Foremost, this book's production was shaped by the protests that started brewing in Hong Kong in the summer of 2019 and that escalated in a manner none of us had foreseen. Along with the protests, geopolitical conflict between the United States and China was building up. If geopolitics could move tectonic plates, Hong Kong had become the epicenter of the shocks. It was in this context that some of the contributors to this book were disinvited from the initial workshop taking place across the border in Shenzhen, while others experienced delays of paperwork that made visa applications unworkable. Of course, the interruption of our workshop only underscores the importance of academic exchange and unhindered discussion in times of tension and conflict. In this regard, we would like to thank the School of Journalism and Communication at the Chinese University of Hong Kong, where the workshop was partially relocated, and especially Francis Lee, Saskia Witteborn, and Jack Qiu, for their ad hoc hospitality and unwavering support for academic exchange.

Another crisis that has shaped this collection is the pandemic of COVID-19. The crisis struck relatively early in Hong Kong, though it thankfully never escalated. Yet collaborating across borders during the pandemic gave us a direct sense of the uneven and unequal ways in which a global crisis strikes. We would especially like to thank Stephen Hartnett, who as the series editor has shown extraordinary dedication to this publication's success despite chaotic times. Likewise, we would like to thank Catherine Cocks at Michigan State University (MSU) Press, for her support and help. It has been a great pleasure to work with her and MSU Press. We thank the anonymous reviewers for their insightful comments on the manuscript. Given the hardships experienced during the pandemic, we would like to thank all the contributing authors for their perseverance.

Rolien Hoyng thanks her partner Murat and her daughter Ruya, who was born in the process of this book's production and who makes the future so much more present and felt, through a mixture of fear and wonder. She thanks her friends and especially her family far away, who have not allowed distance to set them apart. She thanks Tante Maribel for her care for Ruya. Research for this project was supported by Hong Kong's Research Grants Council (RGC project number 23601417).

Gladys Pak Lei Chong thanks her two children—Lucius and Leonard—for their energy, passion, curiosity, and bright spirits, which have given her hope and strength in difficult times. She would also like to thank her mother, sister, and dear friends for their continuous support and trust in the past years. Finally, she wants to thank the Research Grants Council (RGC project number 12610118) and Hong Kong Baptist University for their support. They make fieldwork research in China possible.

Introduction

New, Old, and Uncertain Futures

The notion of "multipolar" innovation was promoted by the World Intellectual Property Organization (WIPO) of the United Nations in 2009, in response to the increase in patent applications from Northeast Asia.[1] The phrase alludes to the fact that, next to Silicon Valley, other major centers of innovation have emerged within Asia, such as China's Shenzhen High-Tech Park, Korea's Pangyo Techno Valley, and India's IT City Bangalore. More evocatively, the notion substitutes concerns over digital divides and exclusion with a promise of worldwide participation in the radical transformation that sounds through slogans such as big data revolution,[2] smart world revolution, and the Fourth Industrial Revolution.[3] However, from a critical angle, what does such multipolarity encompass? What new social orders and socio-technical trajectories of development does it enable? Or what "old" patterns might still be in place?

This edited volume focuses on communication innovation, namely, the shifting ways communication and social organization are mediated by changing designs of infrastructures and platforms. It investigates multipolar innovation communication by mapping the "new," "old," and "uncertain" futures it invokes and produces across geographical contexts. Chasing "path-breaking" and "disruptive" newness might merely set us heading for "old" futures, inscribed with the power relations that mark the present.[4] Yet, to echo Arturo Escobar, can design and innovation be disconnected from "old," unsustainable, and future-canceling practices and ambitions?[5] Can we recover our ability to imagine other futures and quit the conditions that eliminate and foreclose them?[6] Such imaginative capacity negotiates conditions—economic, geopolitical, sociocultural, and ecological—rather than reproducing them under the pretext of breaking with the present.

We investigate communication innovation at a moment when Silicon Valley's dominant role in conjuring and "patenting" technological futures is challenged. This development calls for a comparative approach to communication innovation that maps similarities and differences—or, as we will explain, dynamics of integration and differentiation in communication innovation—across national boundaries and regional affiliations. Accompanied by a good deal of futuristic Sinological orientalism, the Chinese case has become emblematic of multipolar innovation and technological developments that keep intriguing observers for apparently diverting from Silicon Valley's models. For instance, the growth of the Chinese search engine Baidu became possible in the wake of Google's decision to shut down its operation at least temporarily in China in 2010, according to the company, to avoid compliance with censorship and vulnerability to hacks. In retrospect, withdrawal helped China to grow its own corporations, aiding Chinese data sovereignty and technological independence, though transnational financial investments have always continued.[7] Contrary to narratives about Silicon Valley's market-driven breakthroughs, the success of the Chinese platforms BAT (Baidu, Alibaba, and Tencent) owes much to protectionism, their close ties to government, and their uptake of active roles in governing the population.[8] The suspension of Ant Group's IPO in 2020 and the antitrust investigation of online platforms—first Alibaba in 2021, followed by Pinduoduo, Meituan, and other e-commerce platforms—have once again demonstrated the Chinese state's controlling role in stimulating as well as curbing communication innovation. The particularity of Chinese communication innovation has led scholars to ask whether, after socialism and neoliberalism "with Chinese characteristics," we now are witnessing the rise of a platform society "with Chinese characteristics."[9] Guobin Yang proposes the concept of "state-sponsored platformization" to elucidate this specific process of platformatization, which resembles the state corporatist model but also demonstrates technological and market logics.[10]

Yet, though often considered an exception and anomaly within global trends in communication innovation, Chinese platforms seem to partake in, or lead, a broader tendency toward correlating digital infrastructure and innovation with territorial sovereignty, rather than disentangling them. With the Snowden revelations about the global surveillance activities of the National Security Agency (NSA) not yet forgotten, European states prove to be less willing to accept the central position of the United States in global digital networks. Taking place in the context of the European Court of Justice's decision to overhaul the Privacy Shield arrangement—namely, the data-sharing agreement between the EU and the United States—a recent proposal for European data sovereignty contends that European users' data should be stored locally, and it expresses the political will to search for other options. So far backed by Germany and France, project Gaia-X would be "an enabler for platforms 'Made in Europe'—platforms where the potential of A.I. [Artificial Intelligence] can be tapped while privacy is safeguarded, all without reliance on foreign services."[11]

As the examples cited here indicate, while scrutinizing patterns of similarity/difference or integration/differentiation in communication innovation, this edited volume addresses not just the particularity of Chinese vis-à-vis American innovation, but the broader question of a shifting world order and trends that go beyond China. That is, we unpack communication innovation in a world where China has a strong influence by looking at other places in addition, ranging from Ghana to Turkey and Europe. In doing so, we uncover broader trends such as capitalist de-westernization, nascent China-led globalization, and intra-imperialist struggle.

We embark on a critique of communication innovation at times of increased global connectedness *and* antagonism.[12] Whereas "multipolar" innovation at least initially promised global exchange and inclusion, it takes place against the backdrop of intensifying geopolitical tension, whereby digital communication infrastructures no longer serve as the hallmarks of cyber-themed cosmopolitanisms but have become frequent targets of suspicion and sabotage. Most prominently, the US government under Trump has gone to great lengths to convince the public and its allies of the dangers of Chinese innovation, including the digital infrastructure developed by Chinese companies such as Huawei, and social media such as the by then most-valuable start-up TikTok (Douyin inside China, both owned by China's ByteDance Ltd.).[13] With global markets no longer automatically at the disposal of American companies, the Trump administration has portrayed Huawei as nothing less than a PLA-devised weapon, a Trojan horse meant to render America's communication susceptible to Chinese interference.[14] At the time of this writing, it remains unclear what specific course Biden and his government, as Trump's successor, will take. Meanwhile, in the United States, UK, and Northern Europe, communication infrastructures have become the unlikely targets of violent attacks, inspired by online conspiracy theories insinuating that 5G towers associated with Huawei are responsible for spreading the Coronavirus (COVID-19). In the rather different context of Hong Kong, Chinese technological expansion has been perceived with suspicion too. Tensions over the territory's political autonomy have intensified since the Umbrella Movement of 2014. When protesters partaking in the 2019 movement discovered that Mainland China's Guangdong Province intended to extend the Chinese social credit system to Hong Kong, they dismantled existing smart-city infrastructures in an attempt to discover and examine undisclosed functions. A year later, the newly implemented National Security Law alarmed many locals in Hong Kong and moved them to protect their digital privacy by using pseudonyms online and deleting applications, especially if they belong to Chinese companies.[15]

These recent developments go to show that communication innovation can facilitate not only new forms of alignment and affiliation, but also geopolitical tensions and indeed frightening regimes of surveillance and repression. In the context of these developments, the question is whether and how struggles around communication-related rights occur in different places.[16] The suspicion and subversive acts of sabotage against communication infrastructures across different geographies indicate the global breakdown of communication and consensus. They suggest the decoupling of innovation from beliefs in shared futures and trajectories of change as well as shared norms and values related to communication. The Chinese social credit systems appear dystopian in the Western press, but undeniably enjoys rather high approval rates in China itself.[17] In the United States, Edward Snowden continues to be charged with violating the Espionage Act, but he has long been considered a hero in Europe and his statue has traveled to many of the Continent's major public squares. Developing a comparative approach, this book unpacks the politics, ethics, and struggles of multipolar communication innovation and tracks how different formations lead to both hope and fear. Across our case studies, the book argues that communication innovation lies at the heart of bilateral debates between the United States and China and also of international agendas and struggles that overlap with and sometimes contradict existing US and Chinese investments and histories. In this sense, our book offers a truly global examination of how communication innovation impacts our daily lives, political identities, and capacity to imagine and construct futures. The rest of this introduction offers three critical lenses pivoting around the dyads change/continuity, disruption/structure, and integration/differentiation. These lenses can be applied to the overarching themes that cover the three parts of this book: formal innovation, everyday inventiveness, and novelty as technodiversity.

Old Futures: Change/Continuity

Innovation in Western contexts typically denies the historicity of its own material formations, practices, and imaginaries.[18] Its proponents enshrine innovation in an aura of newness, for instance through the incontrovertible seriality of gadgets such as iPhone models that are numerically labeled in ascending order from 1 to n. Yet though imagined, lived, and marketed as novelty, communication innovation often remains contained and embedded in power structures.[19] This enmeshing of the "new" within old power structures follows from the fact that innovation as a process is managed by exclusive institutions and often nation-states, whereas, as material technology, it is inscribed with sociocultural and geopolitical hierarchies.[20] What counts in terms of the critical analysis inquiring into the interplay of "old" and "new" futures are the wider social *effects* and ramifications of innovation and the extent to which they shape societies anew. For instance, data analytics supposedly produce a "new" gaze onto society that focuses on actual behavior rather than assumed identity, and this allows for innovation that "disrupts" industries, markets, and societies. Nonetheless, current technology engenders continuous structural disempowerment, discrimination, and racial profiling, as demonstrated by applications in China that identify Uyghurs specifically, and in the United States that profile African Americans through proxies that can be reduced to race.[21] Neither of these applications disrupts power relations nor do they stir technological instrumentalization away from histories of surveillance and repression of minorities.

Deploying the dyad change/continuity as a critical lens, we raise the question of whether multipolar communication innovation renders redundant geographically oriented critiques of capitalist modes of production. Such critiques address, for instance, the international division of labor, which underscores the geographical distribution of high-skilled and low-skilled labor, and extractivism, which marks processes of dispossession and primitive accumulation of local resources by global players.[22] Nowadays, the capacity for innovation requires access to what is dubbed the most important "raw material" of our times, namely, data that are mined and extracted. What is mined ultimately are our social relations, our private selves, collective behavioral patterns in cities, the logistical flows of goods, and the bioinformatic consistencies of our bodies.[23] Some have argued that data-driven innovation amounts to a global regime of data colonialism that renders everything and everyone a resource for its own reproduction.[24] Yet this perspective fails to acknowledge how unequal levels of disenfranchisement and subaltern status intersect with the power inherent in data and datafication; and more so, that the labor in the process of data-driven innovation still registers particular geographies and social orders.[25] Despite the fact that such geographies have become more complex than simple schemes of First/Second/Third World or Global South/North purport, new types of digital sweatshop labor involve work that machines currently cannot perform either as well or as cheaply as their human counterparts can.[26] Human workers are responsible for image recognition assignments via Amazon Turk, "gold mining," and removal of impermissible content from platforms, in addition to infrastructural maintenance in data centers. What these examples suggest is that communication innovation often draws on, and reproduces, persistent power relations and social orders, which outline particular, though shifting, geographical distributions and ethnic relations.

Uncertain Futures: Disruption/Structure

Next to change/continuity, the second critical lens we introduce revolves around the dyad disruption/structure. Disruption and destruction play prominent roles in mainstream innovation discourse. Joseph Schumpeter's notion of creative destruction naturalizes a logic of capitalism, in which capitalism is spurred by innovation, which allows for creating a temporary monopoly in a new market, while destroying existing industries and institutions.[27] Such effects of innovation are disruptive but also systemic and even exploited as opportunity by the entrepreneurial agents of innovation. Nowadays, investors as well as tech companies themselves speculate in entrepreneurial manner when they invest in, or acquire, promising start-ups in order to capture the next innovation and dominate its market.[28]

Rooted in such a Schumpeterian appetite for disruptive innovations conquering markets, the financialization of innovation has taken on new proportions since the 1980s. Smart-city development, for instance, has proven far from immediately successful, but enterprises have been sustained through speculative bets on innovative potential by investors and shareholders.[29] What matters hereby is not just the immediate success of a particular product or service in terms of technological functionality and social adaption, but the future promise that the company's potential will disrupt markets. The investor bets on the capability of a company to develop new technologies by buying its shares, while hedge funds create markets around the risk of failure. However, the debate about whether financialization encourages innovation or undermines it is divided, with some arguing that financialization happens at the expense of more open-ended research and development activities. Even though it is true that operating special innovation units can enhance a company's reputation and entice investors, resources are not allocated to fundamental research that takes a longer time, or to research that has less market potential.[30] Meanwhile, companies take financial logics to the heart of their corporate decision-making and budget strategies when they spend their profits on buying back their own shares to manipulate prices rather than reinvesting that capital in research.[31]

Following such dynamics of financialization, entrepreneurial activity may in fact limit human and technological potential for creating new futures.[32] However, it should be noted that even though communication innovation may not produce the path-breaking futures that it promises and often remains embedded in structural relations, the effects of technological change are often neither controlled nor foreseeable. They exist as unaccounted-for, and often invisibilized, disruptions, destructions, and risks—in other words, as uncertain and precarious futures. For instance, financialization comes with unequal distributions not only of (potential) profit but also risk. Substantial risk is borne by the Uber driver in India, who invests in a new car but then suddenly faces a decrease in payment when Uber adjusts the pay scale in response to pressure from investors to show a profit. In a secondary cycle of financialization, the option the driver is left with is to apply for a loan, again from Uber.[33] For this driver, the path of securing a better future is full of risk, uncertainty, and potential disruption to their livelihood. Indeed, what Schumpeter's account of creative destruction leaves out are the social costs of this logic, which Marx defined before him in terms of continuous insecurity for labor.

"Uncertain" futures can be understood in terms of the imminent risks of ecological breakdown and catastrophe, indicating the unsustainability of the present. The philosopher Paul Virilio dramatically proclaimed: "When you invent the ship, you also invent the shipwreck; when you invent the plane you also invent the plane crash; and when you invent electricity, you invent electrocution.... Every technology carries its own negativity, which is invented at the same time as technical progress."[34] Virilio's quotation orients us to the destructive nature of innovation. Waste is intrinsically related to innovation when we consider the role of planned obsolescence in, for instance, consumer electronics. It forms an externality of the innovation-driven economy that causes harm and suffering, which often do not appear in any calculation of costs. Risk pertains to the unpredictable environmental and health consequences of innovations such as plasma screens when they become waste and are (illegally) exported to poorer and less regulated regions. Just when environmental regulation and advocacy have forced companies to ban or reduce one harmful component used in electronic devices, the next component is introduced in the name of novelty and innovation, while its environmental consequences remain unknown.[35] In such cases, innovation induces moments of openness and opportunity, but also uncertainty, risk, and destruction.

Multipolarity: Integration/Differentiation

As argued so far, the paradoxes of *change/continuity* and *disruption/structure* are central to our "critique of the new." But how do these themes play out in the case of multipolar communication innovation? Mainstream innovation studies often render context implicit, and such decontextualization results in universalist accounts, which combine celebrations of "path-breaking" disruptive change with narratives that cast technological development as an inevitable, irresistible, and rational movement, unfolding in universal and homogeneous time. The school of diffusionism, which emerged around the middle of the nineteenth century, subscribed to "the idea of technology as historical grand narrative, as a primary determinant of history itself."[36] This school has held that technologies were conceived and created in Europe and subsequently "diffused to the rest of the world almost entirely through European agency and without significant local input."[37] Reiterating aspects of the diffusionist argument more recently, Everett Rogers's much-cited work presents a model for adoption rates that considers technological diffusion a matter of rational choice to adopt or reject a new technology.[38] He divides global society into groups of "innovators," "early adopters," the "early" and "late majorities," and "laggards," who each make their decisions on the basis of knowledge available to them.

Whereas such temporalized discourse renders innovation a matter of universal rationality and singular, ultimately irresistible development, the Global South appears as nothing but an "ontological designing consequence" of the North, at the expense of recognizing context-specific questions, problems, and practices related to design and innovation.[39] In contrast, postcolonial studies has called for "provincializing Europe" to take into account the existence of alternative modernities and perspectives from the so-called "Third World."⁴⁰ For the study of science and technology, this perspective offers analytical tools to decenter West-centric technoscience, while recognizing "hybridities, borderlands and in-between conditions" that reveal other and counter-hegemonic experiences and socio-technical realities.⁴¹ Postcolonial approaches have spurred regional and local social studies of science and technology in, for instance, India, Singapore, Taiwan, and Japan.⁴² Such endeavors at times deploy cultural studies techniques of "inter-Asia referencing" and "Asia as method" to trace similar experiences across Asia and strengthen local agency and solidarity.⁴³

Informed by postcolonial perspectives, this edited volume inquires into today's multipolar communication innovation. Does multipolar innovation imply a continuation of technoscientific universalism or does it enable technodiversity—that is, the emergence of technological, or in fact socio-technical, difference and alternativity?⁴⁴ While the proffered arrival of multipolar innovation suggests global participation in communication innovation, the notion of multipolar innovation does little to challenge the diffusionist logic and temporality. What we can witness nowadays in China is an emerging form of technoscientific nationalism that is built on the historical experiences of technoscience and modernity.⁴⁵ China is not alone, as other East Asian countries have embarked on similar races. Hence, even as an effort to address a globalizing innovation development, the setup of the multipolar model reveals not only globally shifting power relations but also, implicitly, the continuation of dynamics and ideological frames constructing progress, development, and modernity in ways that both seduce and force those "lagging behind" to commit to "catching up."⁴⁶ The recent surge in innovation among the East Asian countries continues this endless loop of "catching up," again erasing actual experiences of disruption, destruction, and harm that are concomitant with being implicated in technological makeover as well as alternative socio-technical realities and possibilities.

This becomes clear when looking at tech companies from more "developed" countries within Asia that are exploring their regional footholds to expand their market share by leveraging innovativeness as competitive advantage, along with geographical and cultural proximity. For instance, in Korea, US-based multinational tech companies, such as Apple and Google, have little presence, while local companies such as Kakao, Samsung, and LG dominate the market. Kakao is a South Korean mobile messaging provider whose shares are partly owned by China's Tencent. It has expanded its operations to include financial services (KakaoPay,

KakaoBank), geolocation services (Kakao T, KakaoBus), and games (Kakao Games). Beyond Korea, KakaoTalk operates in Indonesia, Japan, and Vietnam. There is a need to come to terms with not just waning Western hegemony, but the new territorial divisions of an emerging multipolar world, including the rise of an upper-case "Asia" that dominates, controls, and subordinates the marginals, who are once more "lagging behind."[47] Globally, the development toward "multipolar" innovation signifies a process of capitalist de-westernization by the proverbial "rest," such as the BRICS (Brazil, Russia, India, China, South Africa) countries. Forming sizable blocks that counter US hegemony, their surge does not undermine capitalism and imperialism as much as introduce intra-imperialist struggle.[48]

Though inspired by postcolonial approaches, our endeavor is not to emphasize particularity per se, be it of the institutions of modernity itself or of postcolonial geographies cast as sites of radical resistance and alternativity.[49] Rather, along with paradoxes of *change/continuity* and *disruption/structure*, we aim to underscore dynamics of *global integration* and *differentiation*, as two tendencies unfolding as part of the same movement. For instance, Chinese innovations such as the social credit system are often discussed in the Western press as if they were isolated and unique to China, accompanied by Cold War rhetoric. Yet social credit systems share features with American consumer credit technologies as well as rating mechanisms on digital platforms such as Uber and eBay, and they find an uncanny counterpart in students' surveillance systems operationalized by what are supposedly the very beacons of liberalism, namely, US universities.[50] This example goes to show that any comparative approach should not just relinquish West-centric universalisms relegating others to the past, but also what seems just as pertinent nowadays: cyber-orientalism propelling others into a (dystopian) future at the expense of recognizing mutual implication in technological development. Common technologies and infrastructures are at work, though critical differences exist with regard to their applications and current state of integration across them.

Moreover, as an analytical lens, *integration/differentiation* offers distinct advantages to the endeavor of comparing experiences of innovation and technological development in different contexts. Within the anthropology of technology, the opposites of universalist diffusionism and particularist, culturalist approaches were negotiated by André Leroi-Gourhan, a student of Marcel Mauss, who explored technologies adapting and being adapted to the local milieu in the process of technological evolution. The encounter between new technologies and the particularity of the milieu into which they were integrated can form an instance of invention, but it also conditions and limits the possibilities of technological development.[51] Leroi-Gourhan's point was not to underscore the specificity or "genius" of particular ethnic cultures but to understand processes of technological evolution manifesting itself through diffraction and differentiation. In this volume, we emphasize exactly such processes: the global integration of communication infrastructures and our shared implication in them, along with the heterogeneity of situated concretization, adaptations, and risky ramifications.

Following the dyad *integration/differentiation*, we develop a comparative approach that underscores the ways infrastructures of communication innovation both affiliate us *and* set us apart, and how they implicate us in similar technologies and techniques but also expose us to unwieldy and context-specific adaptations, effects, and ramifications. Our comparative approach builds on the insight that finding similarity opens the way to the discovery of further difference, whereas difference can only become apparent and meaningful against an interpretation of commonality or equivalence at some level, too. This is to say that similarity and difference exist in a symbiotic relation.[52] Sensitizing ourselves to this mutual enmeshment between similarity and difference forms a way of addressing movements of *integration/differentiation* in technological development and creating analytical and normative lenses that lock us to the pole of neither universality nor particularity.[53]

Three Themes

This book is structured around three themes. The first theme explores *formal innovation*, including institutional discourses of innovation, law, political economy, and geopolitics. This theme discusses the planning and regulation of innovation by states or other institutional actors such as the EU. Authors attend to contradictions or coalescences between state and market forces as well as to the disjunction between instrumentalization of planned innovation and unintended and disruptive effects. The second theme considers *everyday inventiveness*, namely, the shared capacity to create, solve, and collaborate, which can challenge capitalism but also is exploited by it. The third theme addresses *novelty as technodiversity*, which encompasses the search for alternative socio-technical, or even bio-socio-technical worlds.

Formal Innovation

Multipolar communication innovation signifies capitalist de-westernization and intra-imperialist struggle. Nonetheless, its agents may not simply copy capitalism or imperialism, but also change where they, or more generally globalization, are headed.

This holds for the futures of platform capitalism and platform imperialism, which seem less homogeneous than assumed in terms of relations between platforms, industry, and state.[54] Dealing with such questions of formal innovation, Lianrui Jia and David Nieborg consider Chinese platforms at the intersection of infrastructure, geopolitics, and finance. As Chinese platforms have become the infrastructures of life and labor in general, their ability to enhance datafication facilitates governance of the population, be it through fintech applications, social credit scoring, or AI-driven judicial processes. Such datafication processes advance financialization of society as much as authoritarian social governance. However, such "indigenous innovation," which is promoted and protected by the Chinese state, does not easily align with aspirations to operate in markets abroad as applications so far have not proven to be "as globally exportable as the platforms and apps coming out of Silicon Valley." Though highlighting the particularity of Chinese platforms as they are integrated with the governance of the population, this chapter forms a very necessary warning against taking for granted the national scale of Chinese communication infrastructure at the expense of underscoring global infrastructural and financial connections and entanglements. As Jia and Nieborg point out, Chinese digital platforms are "deeply plugged into global circuits and networks of financial elites through fundraising, investment, and corporate management." Mapping such networks undermines the narrative of a Cold War type of competition between two hegemons.

In the following chapter, Angela Daly discusses the legal regulation of digital data in the context of multipolar innovation and its geopolitics. In the chapter aptly titled "Neoliberal Business as Usual or Post-Surveillance Capitalism with European Characteristics?," Daly takes the European General Data Protection Regulation (GDPR) as a case study to explore the EU's role and impact as a regulatory power in data protection and privacy. The question is whether the GDPR truly manages to safeguard user data from surveillance capitalism and thereby indicates a turn away from the tendency toward deregulation that has marked neoliberalism. Alternatively, the regulation represents a compromise that sets some boundaries to the operations of Big Tech, but that does not undermine surveillance capitalism in the process, possibly instead stimulating European industries to lead in a (somewhat more) privacy-aware innovation that complies with the GDPR. Scrutinizing the extraterritorial effects of the GDPR as well as the strategies of US and Chinese companies operating within the EU, Daly teases out the nuances and contradictions of the EU's attempt at acting as a regulatory power shaping markets and industries in the context of multipolar innovation.

In the next chapter, Serra Sezgin and Mutlu Binark discuss the tensions between "local" and "global" innovation in the case of Turkey. The Turkish state considers digital games both a potential technology of governance of the population as well as, when exported abroad, a tool for international diplomacy and nation branding. Hence "local" games, grounded in the state's "own" culture and history, are supposed not to merely offer entertainment but to be useful in sectors such as defense, health, and education, along with nation branding. However, by means of a discourse analysis of interviews with game developers in Ankara, Sezgin and Binark argue that these workers undermine the state's framing of indigenous innovation. Turkish game developers think of themselves as members of a global, creative community of game enthusiasts, who leverage a purely individual creative potential to compete in global game industries. Exploring the contradictions between the two sets of discourses, Sezgin and Binark note that the highly individualized notion of creativity that game developers cultivate dampens their resistance to the illiberal cultural milieu in Turkey. Allowing for fruitful comparison with the case of China, this chapter opens up questions about whether liberal freedoms are a precondition for the flourishing of innovation.

Everyday Inventiveness

Next to formal innovation, there is the inventiveness of everyday life, often associated with places where systemic breakdown and decay require people to have certain skills to engage in making their cities livable.[55] This inventiveness again appears in accounts of "pirate modernity," where people have access to new technologies and products thanks to informal production and distribution channels that weaken boundaries between users and producers.[56] Pirate modernities revolve around co-creation practices of imitation and invention. They render the locus, or origin, of innovation ambiguous and hence challenge myths of genius, individuality, and autonomy that undergird the intellectual property regimes of formal innovation.[57] However, recent developments have blurred the boundaries between piracy and formal innovation. For instance, the "Silicon Valley of China," Shenzhen, was long cast as a pirate enclave, derided for lacking originality, before it became celebrated as a space of innovation, creativity, and design.[58] Shenzhen's emergence as a technology hub draws on practices of design and manufacturing infamously known as *shanzhai*, originally a derogatory term in Cantonese to describe cheap knockoffs.

Daniel H. Mutibwa and Bingqing Xia explore the current hype of Maker culture in China, which has emerged since the 2000s from the *shanzhai* culture in Shenzhen and the Pearl River Delta.[59] Engaging with current debates of global Maker culture, the authors discuss the extent to which the framings of *making* reflect "countercultural" values in the context of China's technological development. The analysis is built on a wide array of documentary evidence and an ethnographic study of four makerspaces and hardware entrepreneurial hubs in Shenzhen. It investigates questions such as: What does *making* in Shenzhen reveal about the identities and composition of its digital fabrication communities? In which ways do the aspirations and motivations of these communities reflect countercultural values? Where countercultural values are discernible, how are they reconciled with entrepreneurial motivations and institutional agendas to achieve change? The authors argue that despite the authorities' instrumental (mis)appropriation of countercultural values for its politico-economic ambitions, and the tensions and contradictions within this multifaceted development, *making* practices in Shenzhen carry an open-source ethos and transformative capacity offering makers autonomy for peer production and social intervention. *Making* in Shenzhen does correspond to the grassroots countercultural values of the globalizing Maker movement.

Jian Lin and Jeroen de Kloet explore how the inventiveness of everyday life intersects with the state-commerce relationship through a case study of Kuaishou, an algorithm-based video-sharing platform targeting second- and third-tier Chinese cities as well as the countryside. While existing studies have exposed how the platform economy has contributed to the deterioration of labor conditions, turning individuals into "subcontractors" and "prosumers" without stable wages or benefits, Lin and de Kloet pinpoint how this could overlook the active agency and creative practices initiated by individuals—in their study, the often forgotten, unnoticed, and unlikely "grassroots" (caogen 草根) content producers. These grassroots digital entrepreneurs find their opportunities in social media platforms like Kuaishou. Kuaishou's very existence is closely linked to national policies—"Mass

Entrepreneurship and Innovation" and "Internet+"—and it is firmly in line with the state's order for censorship and social stability. The complicated state-platform relationship distinguishes the Chinese platformization of cultural production from that in the West. Lin and de Kloet argue that institutional regulations and censorship have not stopped these "unlikely" grassroots creators from being creative; more intriguingly, their study demonstrates how these individuals appropriated the algorithmic digital system and negotiated with the state-platform governance to reach their creative and financial objectives.

Novelty as Technodiversity

If we understand technodiversity to imply a disruption of power relations, the question emerges: under what conditions could communication innovation call forth alternative communicative and organizational possibilities in support of social justice? Ruha Benjamin evaluates several initiatives that stage design for social good. She quotes a definition of "design justice" that describes it as "a field of theory and practice" concerned with procedural and distributive justice, namely, with advancing the participation of marginalized groups in design processes and with interrogating how the design of objects and systems distribute risks, harms, and benefits.[60] Such ideas, though attractive, are not new and go back to participatory design, which several authors addressing postcolonial/decolonial computing have problematized in the light of the inequalities that mark postcolonial settings.[61] Even when committed to design for social good and participatory practice, the danger remains that designing technologies and systems for "others" locks them into assumptions about culture, needs, and desired outcomes. Benjamin questions whether "design-speak" itself might not already imply hierarchies and exclusions, privileging professional designers. Meanwhile, design-speak appeals to a desire for novelty in a way that other "old-fashioned" methods of struggling for social justice often do not. Its promise for newness via design and quick fixes to social problems may simply distract from the need for more radical and comprehensive social imaginaries that challenge our ways of life at large. As Benjamin phrases the confusion, "If design is treated as inherently moving forward, that is, as the solution, have we even agreed upon the problem?"[62]

In her chapter, Miao Lu raises the question whether an alternative design process is possible and indeed experimented with by Chinese mobile-phone vendors catering to so-called "bottom of the pyramid" (BOP) markets that are overlooked and deemed unprofitable by global tech giants. She bases her chapter on fieldwork in Ghana with a Chinese company headquartered in Shenzhen, which booked its original success in the domestic rural market and subsequently grew into the biggest manufacturer of cellphones for the African market. Lu examines how the mobile-phone vendor Transsion Holdings seeks to emulate "indigenous innovation." Transsion's *shanzhai*-like innovation practices reveal the persisting gaps between the Western-based normative design, which is often male-, urban-, and white-oriented, and the actual needs of users from peripheral countries. Such gaps could allow tech producers in the Global South to reimagine the use and design of technology and carve out alternative socio-technical worlds. However, while Transsion might have challenged the hegemonic Global North tech designs, its strong presence and growth in the BOP markets could at the same time turn it into the next dominant—albeit emerging—tech company in specific local contexts. The chapter therefore poses questions about the binary opposition between the Global North and Global South, revealing the fluidity and complexities at stake in the global development of communication innovation.

Along with decolonial epistemologies, ecological and more-than-human philosophies can help us think of novelty in the sense of alternative bio-socio-technical relations. Braidotti and Haraway have advanced an understanding of sustainability that involves becoming aware of actual and possible entanglements with human and nonhuman others, bringing about a creative transformation of the self through such sensibility.[63] Novelty, considered along such lines of sustainability, mutuality, and care, could prompt us to explore ways of communicating and organizing that foreground shared existence and the potential to transform.

But such ethical visions, however inspiring, still require embedding in concrete political context. Monika Halkort's chapter explores technologies that were introduced in the name of sustainability and care, yet that end up effectuating surveillance and neglect. She discusses how bioscientific sensing technologies that monitor marine ecologies in the Mediterranean Sea are repurposed as military technologies to surveil migrants risking their lives to make the crossover to Europe. In the process, mediated practices of sensing engender hierarchies, divisions, inclusions, and exclusions. Whereas marine life is cared for, migrant deaths are naturalized and overlooked, even though the vulnerability of these various forms of life in some ways derives from their interdependency and mutual exposure to histories of colonialism, extractivism, and climate change. Halkort's case study goes to show that technological innovations often consist of adaptations. Moreover, it expands the notion of "multipolar" from a non-anthropocentric perspective by highlighting the multiplicity of nonhuman actors implicated in change and transformation. Exploring the violence concomitant with technical incursions undertaken in the name of human ingenuity and progress, this chapter serves as a critical mirror for ongoing and future projects of globalization and colonization that reproduce such myths for the sake of their own legitimation.

In Conclusion

In the concluding chapter of this book, Jack Linchuan Qiu provocatively posits that media and communication scholarship has "long chased cutting-edge innovations," the latest popular brands as well as "trending concepts, methods, memes, and hashtags."[64] But what actually defines novelty and creativity? Qiu questions whether scholarship often remains in the grasp of a fetish with all things "new" because we still lack sufficient critical distance from the Wall Street–dictated futures envisioned in corporate boardrooms, and from rhetoric staked on lingering US-centrism and Chinese exceptionalism. Qiu encourages us to see futures—in the plural—emerging from unlikely places in the Global South and to practice a genuine multipolarism premised on solidarity.

This book hopes to make a humble contribution in this regard by offering a critical framework regarding multipolar communication innovation in the introduction, followed by a set of seven empirically grounded and analytically rigorous studies that cover various geographies, plus Qiu's concluding reflection. As this introduction has argued, whereas innovation induces moments of openness and opportunity to be exploited by a class of entrepreneurs, others merely face uncertainty, risk, and destruction. Hence, our critical approach to innovation reveals paradoxes of *change/continuity* and *disruption/structure* and distinguishes between "new," "old," and "uncertain" futures. Moreover, the narratives of diffusionism and multipolar innovation alike tend to overlook the plurality of experiences, socio-technical realities, and possibilities pertaining to communication innovation. In contrast, the comparative lens of *global integration/differentiation* highlights how technological integration is concomitant with differentiation: infrastructures of communication innovation both affiliate us *and* set us apart, as similar technologies and techniques often result in rather context-specific adaptations, effects, and ramifications. To render visible practices that are either overlooked, marginalized, or considered illicit by mainstream innovation literature, the term "innovation" requires opening up. Per our framework, we can distinguish between *formal innovation*, which is supported by dominant political, economic, and legal apparatuses; *everyday inventiveness*, which resides in the shared capacity to collaborate and co-create; and *novelty as technodiversity*, which imagines and generates alternative socio-technical, or even bio-socio-technical, worlds.

Multipolar innovation seems to coincide with the decoupling of innovation from beliefs in a universal trajectory of change and universal values. The antagonisms and divisions that proliferate at the side of digital infrastructure reflect contrasting public perceptions, values, and regulations. Amidst intensifying division and antagonism, it becomes harder to imagine how to integrate innovation and social justice. Many have presumed a connection between the cultivation of freedoms in a society and that society's ability to innovate. But what is left of the thesis that innovation requires liberal freedoms? China's authoritarianism has apparently not stood in the way of the success of its innovation industries, measured by dominant indicators such as the amount of intellectual property applications.[65] Despite Big Tech aligning itself with the government, this does not mean that industries find themselves constrained in their ability to innovate. Simultaneously, tech industries in the supposedly "free" world are increasingly showing their dark side. The most renowned Silicon Valley brands have gone as far as cultivating secrecy at the expense of integrity of the US democracy, providing misleading testimonies before Congress in the United States and refusing to testify in person in the British parliament, and signing controversial contracts pertaining to military and medical technology, without knowledge or approval of those employees who are supposed to dedicate their creativity and skills to the endeavors. Coincidentally, in times of multipolar innovation, struggle and resistance take up various forms. Sabotage, as in the aforementioned case of smart-city infrastructure in Hong Kong, is but one form of struggle. Tech workers self-organizing to protest their companies, as happened in the United States, is another. Yet given the global impact of innovation and the connectedness of digital communication infrastructure, what is sorely lacking are more cosmopolitan as well as inclusive institutions and organizations that enable effective regulation of innovation.

A different but related issue is participation in innovation. Currently, the social energy and potential of everyday inventiveness are either criminalized by the intellectual property regimes that underpin formal innovation, or they are exploited. For instance, co-creation of culturally specific content drives the big

American platforms such as Facebook and YouTube and has enabled them to build a global reach—a strategy dubbed platform imperialism.[66] From TikTok to Kuaishou, Chinese platforms are attempting to follow suit nationally and internationally by integrating different subcultures and extracting value from mass innovation. Meanwhile, platforms for all kinds of gig work such as Amazon Turk or Zhaopin, and Uber or Didi exploit everyday inventiveness, local knowledges, and savoir faire. Both the Western discourse on "open" and "free" sharing and the Chinese discourse on "mass innovation" incite co-creation and inventiveness, yet may betray the more radical roots of such ideas, namely, socialist as well as Western-countercultural visions of creativity and participation.[67] Whereas grassroots creativity is alive today, at times overcoming the constraints imposed by mediating platforms, these past ideological visions in fact carried aspirations, such as collectively building another world, that are harder to come by today.[68] Integrating innovation and social justice does not just involve better regulation but also resisting the exploitation and constraints imposed on everyday inventiveness, while recovering such social energy and capacity for participation in world-building and imagining futures.

Across the "old" imperialisms of the West and the emergent technonationalisms and intra-imperialist struggles concomitant with multipolar innovation, what remains rather constant is the belief in progress and "path-breaking" innovation. The ideology of newness obscures the very repetition of marginalization of other (possible) ways of life, the exploitation of everyday inventiveness, as well as extractivism and destruction of ecological commons. But, as decolonial and more-than-human perspectives contend, novelty instead can be sought in sustaining, nurturing, and, in doing so, *reinventing* relations with who and what exist around us. Integrating innovation and social justice hence may be better served by the pursuance of reinventing relations than by the infatuation with newness or "design-speak" that promises quick fixes. This requires not just creativity, but also critique of existing conditions and geopolitical, social, and ecological relations that persist despite supposedly "pathbreaking," "disruptive" innovation, or even because of the latter. As editors, we hope that this volume can bring together, and give voice to, such badly needed critique from various geographical contexts and across geopolitical divides.

NOTES

1. Francis Gurry, "Towards a World of Multi-Polar Innovation," World Intellectual Property Organization, November 30, 2009, https://www.wipo.int; and "Global Innovation Index 2019," World Intellectual Property Organization, July 24, 2019, https://www.wipo.int.
2. Rob Kitchin, The Data Revolution: Big Data, Open Data, Data Infrastructures and Their Consequences (Los Angeles: Sage, 2014).
3. Klaus Schwab, *The Fourth Industrial Revolution* (London: Penguin Books, 2017).
4. Clayton M. Christensen, Michael E. Raynor, and Rory McDonald, "What Is Disruptive Innovation?" *Harvard Business Review* (2015).
5. Arturo Escobar, Designs for the Pluriverse: Radical Interdependence, Autonomy, and the Making of Worlds (Durham, NC: Duke University Press, 2018), 15.
6. Escobar, Designs for the Pluriverse, 16
7. Shing Young Yeo, "Geopolitics of Search: Google versus China?," *Media, Culture & Society* 38, no. 4 (2016): 591–605. See also Lianrui Jia and David Nieborg in this volume.
8. Jeroen de Kloet, Thomas Poell, Zheng Guohua, and Chow Yiu Fai, "The Plaformization of Chinese Society: Infrastructure, Governance, and Practice," *Chinese Journal of Communication* 12, no. 3 (2019): 249–56; Min Jiang, "Internet Companies in China: Dancing between the Party Line and the Bottom Line," *Asie Visions* 47 (January 2012), https://www.ifri.org/en/publications/enotes/asie-visions/internet-companies-china-dancing-between-party-line-and-bottom-line.
9. De Kloet, Poel, Zheng and Chow, "The Plaformization of Chinese Society."
10. Jonathan Unger and Anita Chan, "China, Corporatism, and the East Asian Model," *Australian Journal of Chinese Affairs* 33 (1995): 29–53. Guobin Yang, "Introduction: Social Media and State-Sponsored Platformization," in *Engaging Social Media in China: Platforms, Publics and Production*, ed. Guobin Yang and Wei Wang (East Lansing: Michigan State University Press, 2021), xx.
11. Janosch Delcker and Melissa Heikkilä, "Germany, France Launch Gaia-X Platform in Bid for 'Tech Sovereignty,'" *Politico*, June 4, 2020, https://www.politico.eu/article/germany-france-gaia-x-cloud-platform-eu-tech-sovereignty/. See also Angela Daly in this volume.
12. Patrick Shaou-Whea Dodge, "Communication Convergence and "the Core" for a New Era," in *Communication Convergence in Contemporary China: International Perspectives on Politics, Platforms, and Participation*, ed. Patrick Shaou-Whea Dodge (East Lansing: Michigan State University Press, 2021), ix–xxxii.
13. Jill Disis and Jennifer Hansler, "The United States Is 'Looking At' Banning TikTok and

Other Chinese Social Media Apps, Pompeo Says," *CNN Business*, July 7, 2020; Jessica Bursztynsky, "Huawei Expansion in Western Nations May Be 'a Trojan Horse,' Warns a Top GOP Senator," *CNBC*, June 28, 2019.

14. Jufei Wan and Bryan R. Reckard, "Huawei and the 2019 Cybersecurity Crisis: Sino–US Conflict in the Age of Convergence," in *Communication Convergence in Contemporary China: International Perspectives on Politics, Platforms, and Participation*, ed. Patrick Shaou-Whea Dodge (East Lansing: Michigan State University Press, 2021), 97–126.

15. Karen Chiu and Josh Ye, "Hongkongers, Spooked by Beijing's New National Security Law, Are Scrubbing Their Digital Footprints," *South China Morning Post*, July 7, 2020.

16. For instance, see Payal Arora, "GDPR—a Global Standard? Privacy Futures, Digital Activism and Surveillance Cultures in the Global South," *Surveillance & Society* 17, no. 5 (2019): 717–25.

17. Gladys Pak Lei Chong, "Cashless China: Securitization of Everyday Life through Alipay's Social Credit System—Seasame Credit," *Chinese Journal of Communication* 12, no. 3 (2019): 290–307.

18. Escobar, Designs for the Pluriverse, 15.

19. Warwick Anderson, "Introduction: Postcolonial Technoscience," *Social Studies of Science* 32, no. 5 (2002): 643–58; Warwick Anderson, "Asia as Method in Science and Technology Studies," *East Asian Science, Technology and Society: An International Journal* 6, no. 4 (2012): 445–51; Paula Chakravartty and Mara Mills, "Virtual Roundtable on 'Decolonial Computing,'" *Catalyst: Feminism, Theory, Technoscience* 4, no. 2 (2018): 1–4; Anita Chan, *Networking Peripheries: Technological Futures and the Myth of Digital Universalism* (Cambridge, MA: MIT Press, 2013); Ruha Benjamin, *Race after Technology* (Cambridge: Polity Press, 2019).

20. Andrew Feenberg, *Questioning Technology* (New York: Routledge, 1999); Brian Salter, "Biomedical Innovation and the Geopolitics of Patenting: China and the Struggle for Future Territory," *East Asian Science, Technology and Society: An International Journal* 5, no. 3 (2011): 341–57; Chakravartty and Mills, "Virtual Roundtable on 'Decolonial Computing.'"

21. Charles Rollet, "Hikvision Markets Uyghur Ethnicity Analytics, Now Covers Up," *IPVM*, November 11, 2019, https://ipvm.com/reports/hikvision-uyghur; Benjamin, *Race after Technology*.

22. Toby Miller, "The New International Division of Cultural Labor Revisited," *Icono* 14, no. 2 (2016): 97–121; Sandro Mezzadra and Brett Neilson, *Border as Method, or, the Multiplication of Labor* (Durham, NC: Duke University Press, 2013); Ned Rossiter, *Software, Infrastructure, Labor: A Media Theory of Logistical Nightmares* (New York:

Routledge, 2016); Walter Mignolo and Arturo Escobar, *Globalization and the Decolonial Option* (New York: Routledge, 2010); Chan, *Networking Peripheries*.

23. Sandro Mezzadra and Brett Neilson, "On the Multiple Frontiers of Extraction: Excavating Contemporary Capitalism," *Cultural Studies* 31, no. 2–3 (2017): 185–204.
24. Nick Couldry and Ulises Mejias, "Data Colonialism: Rethinking Big Data's Relation to the Contemporary Subject," *Television & New Media* 20, no. 4 (2019): 336–49.
25. Rolien Hoyng, "From Open Data to 'Grounded Openness': Recursive Politics and Postcolonial Struggle in Hong Kong," *Television & New Media* 22, no. 6 (2020): 703–720.
26. Jack Linchuan Qiu, "The Global Internet," in *Media and Society*, ed. James Curran and David Hesmondhalgh (New York: Bloomsbury, 2018), 3–21; Mezzadra and Neilson, *Border as Method, or, the Multiplication of Labor*.
27. Joseph A. Schumpeter, *Capitalism, Socialism, and Democracy* (New York: Harper Perennial, 1942).
28. Sam Dallyn, "Innovation and Financialization: Unpicking a Close Association," *Ephemera: Theory and Politics in Organization* 11, no. 3 (2011): 289–307.
29. Ilia Antenucci, "Infrastructures of Extraction in the Smart City: Zones, Finance, Platforms in New Town Kolkota," *International Journal of Communication* 15 (2021): 2652–2668; Orit Halpern, Jesse LeCavalier, Nerea Calvillo, Wolfgang Pietsch, "Test-Bed Urbanism," *Public Culture* 25, no. 2 (2013): 272–306.
30. Dallyn, "Innovation and Financialization"; William Lazonick, "Innovative Business Models and Varieties of Capitalism: Financialization of the U.S. Corporation," *Business History Review* 84 (Winter 2010): 675–702; You Soo Lee, Han Sung Kim, and Seo Hwan, "Financialization and Innovation Short-termism in OECD Countries," *Review of Radical Political Economics* 52, no. 2 (2020): 259–86; Nick Srnicek and Alex Williams, *Inventing the Future: Postcapitalism and a World without Work* (New York: Verso Press, 2015).
31. Katrin Hahn, "Innovation in Times of Financialization: Do Future-Oriented Innovation Strategies Suffer? Examples from German Industry," *Research Policy* 48 (2019): 923–35.
32. William E. Connolly, *The Fragility of Things: Self-Organizing Processes, Neoliberal Fantasies, and Democratic Activism* (Durham, NC: Duke University Press, 2013); Nick Srnicek, *Platform Capitalism* (Cambridge: Polity Press, 2017); Srnicek and Williams, *Inventing the Future*.
33. Antenucci, "Infrastructures of Extraction in the Smart City."
34. Paul Virilio, *Politics of the Very Worst* (Cambridge, MA: Semiotext(e), 1999), 89.
35. Sean Cubitt, *Finite Media: Environmental Implications of Digital Technologies* (Durham, NC: Duke University Press, 2017); Rolien Hoyng, "Logistics of the Accident: E-Waste Management in Hong Kong," in *Logistical Asia: The Labour of Making a World Region*,

ed. Brett Neilson, Ned Rossiter, and Ranabir Samaddar (Singapore: Palgrave Macmillan, 2018), 199–220.
36. David Arnold, "Europe, Technology, and Colonialism in the 20th Century," *History and Technology* 21, no. 1 (2005): 92.
37. Arnold, "Europe, Technology, and Colonialism in the 20th Century," 87.
38. Everett M. Rogers, *Diffusion of Innovations* (New York: Free Press of Glencoe, 1962).
39. Fry cited in Escobar, *Designs for the Pluriverse*, 6.
40. Dipesh Chakrabarty, *Provincializing Europe: Postcolonial Thought and Historical Difference* (Princeton, NJ: Princeton University Press, 2000); Stephen John Hartnett, "Alternative Modernities, Postcolonial Colonialism, and Contested Imaginings in and of Tibet," in *Imaging China: Rhetorics of Nationalism in an Age of Globalization*, ed. Stephen J. Hartnett, Lisa B. Keränen, and Donovan Conley (East Lansing: Michigan State University Press, 2017), 91–138.
41. Anderson, "Introduction: Postcolonial Technoscience," 645; Arturo Escobar, *Encountering Development: The Making and Unmaking of the Third World* (Princeton, NJ: Princeton University Press, 1995).
42. Gregory K. Clancey, "The History of Technology in Japan and East Asia," *East Asian Science, Technology and Society: An International Journal* 3, no. 4 (2009): 525–30; Togo Tsukahara, "Introduction (1): Japanese STS in Global, East Asian, and Local Contexts," *East Asian Science, Technology and Society: An International Journal* 3 (2009): 505–9.
43. Beng Huat Chua, "Inter-Asia Referencing and Shifting Frames of Comparison," in *The Social Sciences in the Asian Century*, ed. Carol Johnson, Vera Mackie, and Tessa Morris-Suzuki (Acton: Australian National University, 2015), 67–80; Kuan Hsing Chen, *Asia as Method toward Deimperialization* (Durham, NC: Duke University Press, 2010).
44. Yuk Hui, *Recursivity and Contingency* (London: Rowman and Littlefield, 2019).
45. Jean-Christophe Plantin and Gabriele de Seta, "WeChat as Infrastructure: The Techno-Nationalist Shaping of Chinese Digital Platforms," *Chinese Journal of Communication* 12, no. 3 (2019): 257–73; Anderson, "Asia as Method in Science and Technology Studies."
46. Yoshimi Takeuchi, "Asia as Method," in *What Is Modernity? Writings of Takeuchi Yoshimi*, ed. Richard F. Calichman (New York: Columbia University Press, 2005), 149–66; Mizoguchi Yuzo, *China as Method*, trans. Li Suping, Gong Ying and Xu Tao (Beijing: China Renmin University Press, 1996); Chen, *Asia as Method: Toward Deimperialization*; Margaret Hillenbrand, "Communitarianism, or, How to Build East Asian Theory," *Postcolonial Studies* 13, no. 4 (2010): 317–34.
47. Gladys Pak Lei Chong, Yiu Fai Chow, and Jeroen de Kloet, "Toward Trans-Asia: Projects,

Possibilities, Paradoxes," in *Trans-Asia as Method: Theory and Practices*, ed. Jeroen de Kloet, Yiu Fai Chow, and Gladys Pak Lei Chong (London: Rowman and Littlefield, 2020), 1–23. See also Miao Lu in this volume.

48. Miriyam Aouragh and Paula Chakravartty, "Infrastructures of Empire: Towards a Critical Geopolitics of Media and Information Studies," *Media, Culture & Society* 38, no. 4 (2016): 559–75; Chen, *Asia as Method: Towards Deimperialization*.

49. Bruno Latour, Reassembling the Social: An Introduction to Actor-Network Theory (Oxford: Oxford University Press, 2005).

50. Donncha Marron, *Consumer Credit in the United States: A Sociological Perspective from the 19th Century to the Present*, 1st ed. (New York: Palgrave Macmillan, 2009); Chong, "Cashless China," 290–307; Karen Li Xan Wong and Amy Shields Dobson, "We're Just Data: Exploring China's Social Credit System in Relation to Digital Platform Ratings Cultures in Westernized Democracies," *Global Media and China* 4, no. 2 (2019): 220–32; Daithi Mac Sithigh and Mathias Siems, "The Chinese Social Credit System: A Model for Other Countries?," EUI Department of Law Research Paper No. 2019/01, 2019, https://ssrn.com/abstract=3310085; Drew Harwell, "Colleges Are Turning Students' Phones into Surveillance Machines, Tracking the Locations of Hundreds of Thousands," *Washington Post*, December 24, 2019.

51. Bernard Stiegler, *Technics and Time 1: The Fault of Epimetheus* (Stanford, CA: Stanford University Press, 1998).

52. Georgette Wang and Christine Y. H. Huang, "Comparative Guanxi Research Following the Commensurability/Incommensurability (C/I) Model," in *Advancing Comparative Media and Communication Research* (New York: Routledge, 2017), 94–113.

53. Rolien Hoyng, "The Infrastructural Politics of Liminality," *International Journal of Communication* 15 (2021).

54. Srnicek, *Platform Capitalism*; Dal Yong Jin, "Digital Platform as a Double-Edged Sword: How to Interpret Cultural Flows in the Platform Era," *International Journal of Communication* 11 (2017): 3880–98; Marc Steinberg, *The Platform Economy: How Japan Transformed the Consumer Internet* (Minneapolis: University of Minnesota Press, 2019); Plantin and de Seta, "WeChat as Infrastructure."

55. AbdouMaliq Simone, "People as Infrastructure: Intersecting Fragments in Johannesburg," *Public Culture* 16, no. 3 (2004): 407–29.

56. Ravi Sundaram, *Pirate Modernity: Delhi's Media Urbanism* (New York: Routledge, 2010).

57. Laikwan Pang, Cultural Control and Globalization in Asia: Copyright, Piracy, and Cinema (New York: Routledge, 2006); Laikwan Pang, Creativity and Its Discontents: China's Creative Industries and Intellectual Property Rights Offenses (Durham, NC:

Duke University Press, 2012); Sundaram, Pirate Modernity: Delhi's Media Urbanism.

58. Hallam Stevens, "The Quotidian Labour of High Tech: Innovation and Ordinary Work in Shenzhen," *Science, Technology and Society* 24, no. 2 (2019): 218–36.

59. Silvia Lindtner, *Prototype Nation: China, the Maker Movement, and the Socialist Pitch of Entrepreneurial Living* (Princeton, NJ: Princeton University Press, 2020); Silvia Lindtner, "Hacking with Chinese Characteristics: The Promises of the Maker Movement against China's Manufacturing Culture," *Science, Technology & Human Values* 40, no. 5 (2015): 854–79; Clay Shirky, *Little Rice: Smartphones, Xiaomi, and the Chinese Dream* (New York: Columbia Global Reports, 2015).

60. Costanza-Chock quoted in Benjamin, Race after Technology, 175; Chan, Networking Peripheries; Lily Irani, Chasing Innovation: Making Entrepreneurial Citizens in Modern India (Princeton, NJ: Princeton University Press, 2019).

61. Chakravartty and Mills, "Virtual Roundtable on 'Decolonial Computing'"; Kavita Philip, Lilly Irani, and Paul Dourish, "Postcolonial Computing: A Tactical Survey," *Science, Technology & Human Values* 37, no. 1 (2012): 3–29; Mustafa Ali Syed, "A Brief Introduction to Decolonial Computing," *Crossroads* 22, no. 4 (2016): 16–21.

62. Benjamin, Race after Technology, 180.

63. Rosi Braidotti, "The Ethics of Becoming-Imperceptible," in *Deleuze and Philosophy*, ed. Constantin V. Boundas (Edinburgh: Edinburgh University Press, 2006), 133–60; Donna Haraway, *Staying with the Trouble: Making Kin in the Chthulucene* (Durham, NC: Duke University Press, 2016).

64. See Qiu in this volume.

65. Rebecca MacKinnon, "Liberation Technology: China's 'Networked Authoritarianism,'" *Journal of Democracy* 22, no. 2 (2011): 32–46.

66. Jin, "Digital Platform as a Double-Edged Sword."

67. See Mutibwa and Xia in this volume.

68. See Lin and de Kloet in this volume.

Analyzing Chinese Platform Power

Infrastructure, Finance, and Geopolitics

Lianrui Jia and David Nieborg

In late 2019, against the background of a US-China trade war and an emerging global pandemic, US politicians, pundits, and journalists debated the supposed threat of China. The titles of Judiciary Committee hearings chaired by US Senator Josh Hawley (R-MO) are telling: "How Corporations and Big Tech Leave Our Data Exposed to Criminals, China, and Other Bad Actors" and "Dangerous Partners: Big Tech & Beijing."[1] The rise of TikTok, the video-sharing app developed by the China-based tech company ByteDance, particularly worried critics. The app's "happy-go-lucky rise," journalists for the *Washington Post* reported, "was largely shaped by its Beijing-based parent company, which imposed strict rules on what could appear on the app in keeping with China's restrictive view of acceptable speech."[2] This set Senator Hawley up to introduce bill S.3455, or the "No TikTok on Government Devices Act," in March 2020.[3] Running through these articles and public hearings is the idea that China-based tech companies, including apps such as TikTok, have become a serious threat to US hegemony, if not the very future of the internet.

Depending on one's political leanings and nationality, these dire warnings are either long overdue or the hallmark of hypocrisy. Spurred by China's inability to stem the spread of COVID-19 beyond its borders, those who are instinctively wary of China's global ascendance will undoubtedly feel validated by the Trump administration's nativist tendencies. Then again, the vilification of China by US conservatives and pundits purely on economic grounds is duplicitous at best. After all, US-based platform companies are dominant on an economic, financial, and infrastructural level and have benefited from decades of both direct and indirect state support, such as direct access to finance capital and favorable intellectual property regimes.[4] Moreover, while divergent, the values, norms, and infrastructural ambitions espoused by US platform executives, such as Jack Dorsey and Mark Zuckerberg, are said to run counter to non-US societal norms.[5]

In this chapter we consider the political economy of China-based platforms in a moment of multipolar innovation. Our main focus is on the BAT platforms Baidu, Alibaba, and Tencent, from their pre-initial public offering (IPO) stages up until to 2019. Rather than considering Chinese platform companies as "bad actors" or "dangerous partners," we provide an empirical account of their economic, financial, and infrastructural ascendance, both in their domestic and international markets. Do China-based platforms threaten US hegemony? Closer inspection of key metrics suggests otherwise. For example, while the widely popular app WeChat has over 1.2 billion active users, this pales in comparison to Facebook's global user base of 2.89 billion, or WhatsApp's 2 billion users. ByteDance's Douyin (the domestic version of TikTok) has seen rapid uptake in and outside of China, but in 2019, its domestic success (442 million users) was far greater than TikTok's, which sported 37 million users in the United States.[6] A similar argument can be made when considering revenue and market valuations. Shenzhen-based Tencent and Hangzhou-based Alibaba rank high in the list of public corporations by market capitalization, yet they trail the trillion-dollar valuations of Microsoft, Apple, and Amazon. To account for the emerging yet still diffuse power of China-based platforms, this chapter asks: How is power operationalized by China-based platforms? We ground our analysis in a multilevel conceptualization of platform power outlined below, which allows us to situate platform companies in broader ecosystems and political economies.[7] What emerges from our analysis is a more complex picture of the integration of markets and infrastructures. While companies such as ByteDance, Tencent, and Alibaba are distinctively Chinese—in the sense that they cater primarily to a significant domestic market and are fully integrated with state-sanctioned policy frameworks—this qualifier becomes more muddled considering flows of finance capital and corporate ownership.

Locating Platform Power

Leading platforms in China started out as web companies focusing on one or a few key industry segments. Alibaba was founded in 1999 as an e-commerce company, while Baidu (est. 2000) was founded as a search company. Tencent's (est. 1998) historical roots lie in its online chat program QQ. As market leaders in their distinctive market segments, these companies heavily diversified by expanding into and integrating with "sectoral platforms" that include transportation, health, and education.[8] Because of their expansion and integration with other platforms, it becomes increasingly difficult to untangle their reach. Exacerbating this analytical challenge is the process of "interplatformization"; China-based platforms are much more integrated on both an economic and infrastructural level, allowing users to freely share content across platforms, and therefore have fostered a more profitable environment compared to Amazon, Facebook, and Google.[9] Therefore, rather than analyzing how each of these platforms constitute all-powerful monolithic entities, we follow van Dijck, Nieborg, and Poell, who call for greater specificity in analyzing platform power.[10] To untangle the different institutional dimensions of platform power, we discuss how platform power is operationalized on the infrastructural, financial, and geopolitical levels.

First, *infrastructural* power entails platforms' role as societal infrastructures, both domestically and across different geographical areas where they provide data, internet, and surveillance infrastructures, payments, and logistics. This section draws from recent work on the "platformization" of infrastructure and the "infrastructuralization" of platforms.[11] By broadening the analytical scope, this perspective allows for platform power to be considered holistically as it requires us to look beyond each company and beyond measures of market share and ownership. This includes examining how platforms accrue unfair advantages by controlling specific nodes in integrated platform ecosystems, through gatekeeping, lock-in, cross-subsidizing, or combining crucial data flows. These nodes are understood as "infrastructural platform services," which include social networking services, search engines, app stores, advertising systems, cloud services, and payment systems.[12] Platform companies, on their part, consist of a large number of such services, each of which functions as a market that brings together end-users (consumers) and complementors, such as business actors, advertisers, and government agencies.[13]

Second, platform power has a distinctive financial dimension.[14] Not only do China-based platforms constitute typical "winner-take-all" markets but Chinese platforms also leverage financialization by wiping out competition and consolidating market dominance through mergers and acquisitions. To this end they established investment arms and have embarked on equity investment as a means for growth. This makes them not only market participants, but also financiers, investors, and key stakeholders in the global platform economy. The second level of analysis goes beyond the level of infrastructural platform services and takes the platform *ecosystem* as the unit of analysis. At this level it is not only the accumulation of data, but the strategic deployment of investment capital that allows platform companies to extend beyond their boundaries.

Third, we will situate Chinese instances of platform power within the broader *geopolitical* platform ecosystem. Combining the first and second perspectives, here we consider the various partnerships and cross-appointments on boards of directors between leading Chinese platforms with US-based behemoths such as Google and Amazon. These partnerships have the potential to crowd out competition outside of China, particularly in emerging e-commerce markets such as Southeast Asia. Despite the US-China trade war waged during the Trump administration, we demonstrate that US and Chinese platforms share mutual interests as evidenced by their collaboration in establishing markets and the ability to control global data streams. As such, rather than a radical break, this geopolitical convergence of corporate interests suggests that multipolar innovation in the age of platforms is predicated on, and further deepens, capitalist power structures.

Before we discuss these three institutional dimensions, we first canvass the extant literature on the globalization of Chinese digital platforms and provide an overview of Chinese platforms in the context of China's cyber-power construction and globalization projects, highlighting key state policies, initiatives, and the roles of platforms therein. To conduct our multilevel analysis of Chinese platform power, we rely on annual reports, financial reporting, press releases, as well as reporting in the financial press. Our analysis aims to investigate the "threat of Chinese platforms" narrative through an empirically informed critical political-economic analysis. Similar to their US-based counterparts, Chinese instances of platform power manifest themselves differently across multiple institutional levels, and dispersed through different geographic regions and spheres of influence.

Building the Digital Silk Road

The global diffusion and uptake of China-based platforms is both the outgrowth of the country's long-standing "Media Go Global" policy and the result of the economic imperatives driving the growth of platform markets. The Media Go Global policy is a media-focused framework that involves both state and commercial actors to tackle China's global soft-power deficit.[15] As we discuss more in depth below, platform companies are considered important drivers for innovation in China's domestic market. "The platform economy" in China, Julie Chen contends, "is often associated with the ideas of openness, harmony, or green consumption and by extension a more responsible and sustainable metropolitan lifestyle."[16] CEOs of platform companies, on their part, act as "prophets of mass innovation in China," extolling the virtues of their company's services while hewing closely to state-defined understanding of indigenous innovation policies.[17] For example, next to food delivery, ride-hailing platforms such as DiDi have become both digital utilities for urban transport and a means to employ hundreds of thousands of ex–factory workers.[18] Similarly, popular apps such as Alibaba,[19] WeChat,[20] Kuaishou,[21] and Douyin[22] are deeply integrated into everyday practices of hundreds of millions of Chinese citizens. These examples go to show that Chinese platform companies not only benefit from a significant domestic market but also contribute to a decidedly positive collective framing of its economic and societal impact.

Next to domestic development, research has focused on the converging interests between China's state-led globalization project and those of Chinese digital platforms. The most recent state-led project is the Digital Silk Road, a subset of the Belt and Road Initiative (BRI) formalized in 2013, which aims to build a trade and infrastructure network connecting Asia with Europe and Africa. The Digital Silk Road is considered a "growing and complex alliance" formed between the Chinese state and its homegrown internet companies positioned to advance a broad set of economic and political goals.[23] One of the most visible and active companies in this broader project is China's e-commerce giant Alibaba. China's Belt and Road Initiative has offered a major boost to the company's global expansion, particularly its cloud computing business.[24] Similarly, Alibaba's global trade project—called the Electronic World Trade Platform (eWTP)—runs parallel to the BRI and marks a bold initiative to shape global trade that challenges US hegemony.[25] In other words, the grand project of expanding a Chinese digital

TABLE 1. BAT Revenue Generated outside of China

YEAR	2011	2012	2013	2014	2015	2016	2017	2018
BAIDU	0.4%	0.5%	0.2%	0.5%	N/A	N/A	N/A	N/A
ALIBABA	N/A	18.8%	12%	9.2%	8.5%	7.5%	9%	8%
TENCENT	5.2%	5.2%	8%	8.9%	6.4%	4.9%	3.4%	2.9%

Source: Figures reported in company annual reports.

empire is expedited with the participation of the country's digital giants, who have the capacity and expertise to conduct infrastructural and logistical operations on a regional and global level.[26]

Moving beyond business decisions undertaken by individual platforms and their executives, scholars have sought to measure the degree of internationalization of Chinese internet companies. Yin and Li demonstrate that state ownership or a government affiliation increases the international footprint of state-owned Chinese internet companies.[27] The tradeoff, however, is that they have to forego the short-term goal of making profits. In practice, political clout, visibility, and foreign investments do not necessarily translate into profitability. Chaperoned by the state, platform companies expand globally via highly symbolic launches of services during high-profile diplomatic visits, especially after China allowed private actors to conduct its cyber diplomacy. Then again, in very few instances does the display of political backing match market competitiveness. For example, during President Xi's 2014 visit to Brazil, search-engine giant Baidu launched its Brazilian subsidiary, Busca. Baidu's Latin American strategies also included investments in Peixe Urbano and other regional expansions into Argentina, Chile, and Mexico.[28] Unable to break the monopoly of Google, Baidu shuttered its Brazilian operations in 2018.[29] Similarly, in 2011 Baidu launched the Arabic question-and-answer service Hao 123 in Egypt, only to close it six years later.[30] The company's ventures into Japan, Thailand, and Vietnam have not been successful either, raising questions of whether Chinese tech companies' international ambitions live up to the portrayal of a tech juggernaut.[31]

Nonetheless, Chinese platform executives are explicit about their domestic and global ambitions.[32] In a meeting with former Chinese propaganda chief Li Changchun, Baidu's founder Robin Li stated that the company's goal is to become a universally recognized brand in over half the world's countries.[33] As illustrated in Table 1, the BAT trio derives most of its revenue domestically, and in relative terms, global revenues have seen much slower growth. As noted in our introduction, even the revenue and userbase of the first truly "global" Chinese mobile app, TikTok, trail far behind Facebook and Google's app offerings. For the time being, while Chinese platforms expand into different geographic regions, their global footprint still is relatively limited. Next, we will discuss the three dimensions of Chinese platform power, starting with infrastructural power.

The Platformization of Chinese Infrastructure

Chinese platforms' infrastructural power is as much the result of a capitalistic logic of encapsulating and controlling markets as it is the outcome of the state's techno-nationalism projects and policies, both domestic and abroad. Under the aegis of becoming a cyber superpower, the Chinese government launched several national technological development projects, such as the Social Credit System, the National Artificial Intelligence Development Plan, and the Internet Plus Plan.[34] Leading digital platforms are handpicked by the state to participate in national technology plans as they are well-positioned to support the technological infrastructure for the country's informatization and datafication processes. Meanwhile, abiding by the principle of "cyber sovereignty," the Chinese government is pursuing a proactive role in governing cyberspace through refurbishing state control over online activities and transforming, streamlining, and digitizing the delivery of government services and social control.[35] Leading digital platforms, leveraging their market dominance, are key stakeholders in the design and operation of the platformization of digital infrastructures.[36] Two infrastructural projects stand out: the platformization of payment systems and building data infrastructures that support both a national surveillance infrastructure and a broad range of commercial services.

Two of the most transformative instances of the platformization of Chinese infrastructures are the payment systems provided by Tencent and Alibaba's spinoff Ant Group. In China, their services have been able to proliferate because of the historically low utilization rate of credit cards, the annual tradition of sending so-called red packets during Spring Festival, and the government's support for the "fintech" sector and the promotion of "inclusive finance"—which is the belief that digital financial services and online lending address the issue of financial inclusion into broader Chinese society.[37] In practice, platform-based payment services are predominantly accessed via mobile apps, such as the Alipay app or WeChat Pay (integrated in the WeChat app). When buying physical goods, rather than swiping or tapping a credit card at the point of payment, both apps allow users to scan a vendor-generated QR code for seamless payment. Both apps are prototypical "infrastructural platform services": they are integrated within the broader data infrastructures and platform ecosystems of their parent companies and function as the infrastructural tissue integrating users, vendors, and banks, and also other platform services and stakeholders (i.e., the state).[38] By measures of transaction volume and user penetration, mobile payment is nearly ubiquitous. Payment apps reach 92.4 percent of mobile internet users, and both apps constitute a tight domestic duopoly, with 55.1 percent and 38.9 percent market share respectively as of 2019.[39]

Next to domestic dominance, the Alipay/WeChat duopoly has expanded globally as well, spurred by Chinese tourists and diasporas who are increasingly using payment apps to complete transactions overseas. In an effort to tap into global markets, Alibaba and Tencent have relied on a combination of taking ownership stakes or setting up joint ventures with foreign fintech companies. In 2018, Alipay's parent company, Ant Financial, accounted for a whopping 35 percent of global venture-capital investment in fintech firms.[40] Ant Financial has made a particularly strong push into the Southeast Asia region, one of the geographical foci of China's Belt and Road Initiative, through investments in Thai e-payment services Ascend Money, Philippines-based fintech venture Mynt, the Singapore-based firm M-Daq, Indian mobile-payment provider Paytm, Korean's KakaoPay, and by setting up a joint venture with Indonesia-based Emtek. Beyond Asia, Alipay joined a partnership with payment-processing company First Data and Verifone to expand payment systems to North America.[41] Tencent, on the other hand, has invested in Indonesian Go-Jek—a ride-hailing, logistic, and digital payment company—and launched WeChat Pay in Malaysia, Thailand, and twenty-one other countries. As transactions routed through mobile payment systems often escape taxation, WeChat and Alipay pose problems for national financial regulators. As a result, Nepal banned both apps as their use among tourists and the Chinese diaspora resulted in a loss of the nation's foreign-exchange income and tax avoidance.[42] Later in 2020, Nepal did grant both companies a license after they complied with Nepali central Rastra Bank.[43] In short, the global integration of Chinese payment infrastructures are highly uneven and inherently subject to local regulations and institutional contexts.

Harnessing access to finance capital, both platform companies are transforming financial infrastructures by integrating with platform data infrastructures. Former Alibaba CEO Jack Ma proposed the idea of "TechFin," which is meant to signal a full rebuilding of the financial system with a technology-first approach, as opposed to the more common label "FinTech," where technology's role is to improve the incumbent financial infrastructures.[44] For TechFin-focused platforms, data analytics are considered a key competitive advantage. This involves the collection of financial data as well as developing in-house algorithms, machine learning, and AI technology.[45] In this context, digital platforms are well positioned to accumulate consumer data through integration with other infrastructural platform services—e.g., search, e-commerce, and live-streaming—all of which generate data to be used for automated credit assessments. As such, consumer-facing apps transform financial services in platform-dependent practices.

In addition to processing payments, TechFin platforms have broadened their portfolio of financial services, including loans, investment funds, and crowd-funding. Alipay's offerings include payment, clearance, settlement, and investment.[46] For example, its investment app YuEBao invited users to move money from their debit accounts into its investment fund by offering higher interest rates compared to traditional banks.[47] In 2019, out of Alipay's 700 million users, 588 million invested in YuEBao's fund, which equaled approximately one third of the Chinese population.[48] At that point, YuEBao held the world's third largest market funds, totaling $157 billion.

As China seeks to establish a national database for credit information, the aggregation of financial and transactional data in the hands of just two digital platforms has become an important tool to fill blind spots in its centralized credit-scoring system: The People's Bank of China's (PBOC) Credit Reference Center. According to the PBOC, only about 300 million citizens have enough information on file to generate a credit score. Therefore, in 2015 the PBOC licensed eight platforms, including Tencent Credit and Ant Financial's Sesame Credit services, to form Baihang Zhengxin, a unified national credit platform for online lending. Yet without mandated data sharing, Ant Financial and Tencent have so far resisted sharing personal information and credit data with Baihang Zhengxin, creating hurdles in the implementation of the credit reporting database.[49] The power that rests in Chinese platform companies, as they become proprietors of valuable user data, complicates the ongoing convergence process that aligns technology and business leaders with the Party, with President Xi at the core.[50] The sudden halt of Ant Group's initial public offering in late 2020, and a subsequent anti-monopoly campaign aimed at domestic platform companies marked a dramatic turn of events where the Party exerted direct regulatory control over the digital financial industry.

Societal Data Infrastructures

The platformization of payment infrastructures is part of a broader push towards the construction of a centralized, sovereign, indigenous data infrastructure that includes the active participation of domestic market actors who offer infrastructural platform services that afford data collection. Gruin has pointed to the decidedly authoritarian nature of China's financial system, which comprises an array of big data technologies, financial firms, and financial practices such as digital credit scoring.[51] Arguably one of the more evocative examples of this authoritarian approach has been the construction of the Social Credit System (SCS), a national project that sets a comprehensive outline to establish a data infrastructure for social scoring.[52] Started in 2015, the infrastructural backbone of the SCS is the National Credit Information Sharing Platform (NCISP), which connects 42 central agencies, 32 local governments, and 50 market actors.[53] Leading platforms, such as Alibaba and Baidu, also share data with the NCISP.[54]

As with any infrastructural effort of this scale, these investments have a decidedly material dimension. Chinese platform companies have built a sizable physical computing network that includes data centers and cloud services. Unsurprisingly given their position at the heart of the Chinese platform economy, BAT are the three largest players in the domain of cloud computing, owning 8.8 percent, 46 percent, and 18 percent market share, respectively.[55] Tencent and Alibaba's infrastructure is increasingly integrated with legacy service providers, particularly the nation's telecommunication operator China Telecom.[56] Similar to the global ambitions of its financial services, Tencent and Alibaba openly challenge the market dominance of Google, Microsoft, and Amazon in the Southeast Asian region. So far, Tencent has ten overseas data centers and Alibaba eleven.[57] In 2020, Alibaba announced a $28 billion investment in cloud computing services.[58]

Outside of finance and cloud computing there are ample examples of Alibaba and Tencent engaging in the platformization of social and political practices. One of the more dystopian examples is the assistance provided to local police in a number of state-led "smart city" projects. Platform companies have a second job by assisting city officials to build state surveillance networks and use cloud-based data systems and facial-recognition programs to identify and arrest criminals, and to track and even forecast crowd movements.[59] Alibaba's City Brain, an AI-driven system to decrease traffic congestion and improve the detection of accidents, was implemented in 23 cities across Asia, including Shanghai, Guangzhou, and Hangzhou.[60] With

WeChat's widespread adoption among the Chinese internet population, the CCP started a 26-city trial to replace traditional state-issued social security cards with digital versions tied to WeChat user accounts.[61] To spur wide-scale adoption, the service can be used to register at hotels, purchase train tickets and board flights, apply for government services, and open bank accounts. Combined with China's existing real name registration policy, it is nearly impossible to use utility apps such as WeChat anonymously. Thus, WeChat's identification functionality ensured the app's elevation to the status of a vital digital utility for nearly all Chinese citizens.[62]

Next to identification and financial services, digital platforms have made inroads into digitizing China's legal processes. In 2015, the Supreme People's Court of the People's Republic of China recognized the use of WeChat messages as evidence for civil cases, and the admission of WeChat records without the need for notarization.[63] Twelve provincial courts have tried out "mobile courts," operated through the WeChat Mini Program, which includes technologies such as facial recognition, video conferencing, and digital signatures.[64] In 2017, the City of Hangzhou—where Alibaba is headquartered—launched the first "Internet Court" with Alibaba playing a key design, engineering, and operational role. The court handles cases such as online purchases and disputes, online defamation, domain names, and copyright issues. The Alipay app serves as identity verification, and its e-commerce services—Taobao and T-Mall—provide transaction records as evidence. Alibaba Cloud services, then, provide data encryption, storage, and monitoring. In this case, the government not only benefits from the platform's infrastructural affordances, but also draws on the company's experience in adjudicating online disputes as Taobao has built in dispute resolution mechanisms. As of 2019, there are three "Internet Courts" located in Hangzhou, Beijing, and Guangzhou, collectively processing over 120,000 cases.[65]

All these instances of platformization are indicative of a sustained effort to seamlessly integrate platform infrastructures with legacy systems and social and civil practices. In their reflection on the emergence of a North American and European "Platform Society," van Dijck et al. raise concerns about the blurring of the public and the private, and the integration of platform services in sectors such as news, education, urban transportation, and health care.[66] Already, the level of integration of Chinese platform infrastructures with civil institutions and utilities has reached a level far beyond the legal and normative abilities of Facebook, Google, and Amazon. The Chinese Platform Society is a fait accompli—at least considering the roadmaps provided by the state. In 2017, the State Council issued the Next

Generation Artificial Intelligence Development Plan, in which the government handpicked four domestic tech companies to co-develop artificial intelligence open innovation platforms: Baidu for self-driving cars, Alibaba for smart cities, Tencent for medical imaging, and iFlyTek for voice recognition. The four-company national AI team was later upgraded to fifteen, to further advance and integrate the development of AI in finance, education, health care, and "smart homes."[67]

The Financialization of the Platform Economy

Next to infrastructural power, China-based platform companies leverage access to finance capital to shape market conditions, such as market entry, pricing, and above all, corporate ownership. Since their launch, the BAT platforms benefited from access to foreign investment capital: Baidu received investments from Draper Fisher Jurvetson ePlanet Ventures, Peninsula Capital, Integrity Partners, and Google; Tencent received investment from IDG Capital and Pacific Century Cyberworks, as well as the South African media giant Naspers; Alibaba turned to financing by Yahoo! and SoftBank.[68] The decision to raise funds through public offerings further planted these digital platforms tightly into global circuits of capital and subjected them to the regulatory frameworks of foreign stock exchanges. Meanwhile, bearing much resemblance to their Silicon Valley counterparts, the BAT platforms feature centralized ownership control by its founders. Baidu's CEO Yanhong Li is the company's largest shareholder, owning 16.4 percent of shares through his Handsome Reward Limited company based in the British Virgin Islands. Ma Huateng is the largest shareholder of Tencent, owning 8.58 percent of its shares. Lastly, Alibaba has taken more of a partnership approach, where the 38-member Alibaba Partnership, administered by a five-member partnership committee, retains the exclusive right to nominate and appoint a simple majority of their board of directors.[69]

Financialization strategies shift the role of companies from direct market participants to financiers, owners, and stakeholders in the platform economy. Similar to their infrastructural ambitions, growth and expansion strategies have both a domestic and global dimension, and share the same goal: to establish market dominance. This is most visible in the domestic setting, where the BAT companies acquired 75 percent of all successful start-up companies.[70] Mergers and acquisitions are at a historical high, benefiting from debt financing and resulting in increasingly concentrated markets.[71] Waves of consolidation have created conglomerates of an unprecedented scale and scope; the market capitalization of the BAT trio takes up nearly 97 percent of the market capitalization of all publicly listed Chinese internet companies. Despite their size, there is a jarring disparity between profitability and market capitalization.[72] As of 2019, Alibaba had a market capitalization of $567 billion, approximately 48 times its net income of $11.95 billion, whereas Tencent's market capitalization ($509 billion) was roughly 38 times its net income ($13.42 billion).[73] Comparatively, for Amazon this was 83 times, Alphabet 27 times, and Facebook 32 times.[74]

Because of their deep financial pockets, platform companies have become financiers, investors, and stakeholders in the domestic economy. In 2017, under the Internet Plus initiative and in an effort to revitalize the state-owned telecommunication operator China Unicom, BAT injected $11.7 billion in capital.[75] In 2019, Alibaba poured $8.7 billion of investments into the state-owned mobile communication infrastructure company China Tower Corp.[76] These investments mark unprecedented steps by the CCP as it permits private platforms to finance state-owned enterprises in a push to reform legacy ownership structures.

The financial strategies and business models of leading platform companies have steered towards traditional capitalist market imperatives, such as maintaining stock valuations and maximizing shareholder value.[77] To spur financial growth, BAT have all set up venture capital (VC) units to fund technology start-ups.[78] VCs help Chinese internet companies to stay afloat in turbulent markets, fend off competition through acquisition, and serve as lucrative revenue streams whenever any portfolio company goes public. In 2017, Baidu established Baidu Venture, which focuses on artificial intelligence, one of the core technologies the platform is pursuing. In 2018, its venture fund was one of the world's most active investors in AI when counting the number of deals. Arguably, Tencent has been the most aggressive investor, where the platform devises investment as one of the key strategies for growth. In 2018, Tencent initiated an organizational shakeup and stepped up its investments in the media industries and information and communication technologies.[79] After decades of having no presence in the game industry, Tencent has become the number one game publisher in the world in a matter of years, predominantly through strategic investments and acquisitions.[80]

Table 2 indicates the rise of investment income in Tencent's and Alibaba's total revenue. Notably, in 2016, Alibaba's interest and investment income rose to RMB 52,254 million, and this was due to the deconsolidation of two entities: Alibaba Pictures and Alibaba Health. Tencent, on the other hand, has profited from the

TABLE 2. Income from Investment for Tencent and Alibaba

YEAR	TENCENT		ALIBABA	
	OTHER GAINS, NET (RMB, MILLIONS)	OPERATING PROFIT (RMB, MILLIONS)	INTEREST AND INVESTMENT INCOME, NET (RMB, MILLIONS)	INCOME FROM OPERATIONS (RMB, MILLIONS)
2007	69	1,635		
2008	6.9	3,246		
2009	−58.2	6,020.5		
2010	38.1	9,838.2		
2011	420.8	12,253.6		
2012	−284	15,479.4	258	5,015
2013	904	19,194	39	10,751
2014	2,759	30,542	1,648	24,920
2015	1,886	40,627	9,455	23,135
2016	3,594	5,117	52,254	29,102
2017	20,140	90,302	8,559	48,055
2018	16,714	97,648	30,495	69,314
2019	19,689	118,694	44,106	57,084

Source: Author's compilation of companies' annual reports. Tencent is listed on the Hong Kong Stock Exchange in 2004 and Alibaba is primarily listed on the New York Stock Exchange since 2012; therefore there is a difference in financial accounting standard as regulated by each stock exchange. "Other Gains" denotes "changes in fair values of financial assets held for trading" and includes gains on financial instruments and financial assets, interest income, and government subsidies. For example, Tencent's value gain from the IPO of Meituan Dianping was reported under this category. The spike in 2017 was a result of the IPO of companies Tencent invested in, such as Yixin, Netmarble, Sea, ZhongAn Insurance, and Sogou. Compared to operating profit, which increases steadily over the years, other gains fluctuate and feature more significantly as a revenue stream. Alibaba's net Interest and Investment income consisted of interest income, gain or loss on deemed disposals, disposals and revaluation of long-term equity investments, and impairment of equity investments. Alibaba's gains from the Cainiao Network, Koubei, and Alibaba Pictures are also recognized in this category.

initial public offering (IPO) of two of its subsidiaries: China Literature in 2017 and Tencent Music in 2018. After a decade of receiving foreign investments, Chinese platform companies reached a level of capitalization that allows them to deploy financialization as a growth strategy.

Overall, the financialization of Chinese platforms simultaneously bears similarities and historical specificities. On the one hand, financialization, as a historical transformation of capitalism, is marked by an increase in profit making constituting the spheres of circulation and finance.[81] Chinese digital platforms, being deeply plugged into global circuits and networks of finance through fundraising, investment, and corporate management, are leveraging financialization to sustain profitability, stock valuation, and market capitalization. The financial power wielded by the BAT platforms far surpasses other smaller and middle-sized platform companies in China. This amounts to not only higher barriers to market entry and increased competition, but also decidedly different abilities to generate continuous profit and manage risks. On the other hand, as scholars have shown, financialization proceeds in China in a pragmatic manner: undergirded with datafication processes to advance authoritarian social governance, the Chinese state manages financialization to achieve its developmental goals.[82] For digital platforms in particular, the financialization process is driven both by capitalist imperatives and neoliberal state policies, namely, the promotion of the "share economy," which masks issues of equal participation and revenue distribution under rosy ideas of openness, harmony, and green consumption.[83] Moreover, the call to advance "inclusive finance," which led to the siphoning off of individual savings into private platform companies, further looped non-financial actors and household savings into the financialization process. In this regard, the financialization of Chinese digital platforms is indeed the co-creation of the state and capitalist digital platforms.

Platform innovation in China echoes the dyadic tension between disruption and structure: On the one hand, financial innovation serves to reinforce platform owners' market dominance and helps maintain social stability and enhances the Party's legitimacy. On the other hand, innovation led by private platform companies is disruptive to the socialist principles upheld by the Party, as rampant pursuit of profit has resulted in labor precarity, degradation of consumer welfare, monopolistic competition, and the hollowing out of corporate social responsibilities.[84] These negative externalities alerted the Chinese state to improve its attitude and approach to platform expansion and competition.[85] For example, the *People's Daily* publicly called out and reprimanded platform executives for "excessive" commercialization of online services, and called upon them to aim higher, i.e., focusing on technological innovation instead of short-term profits.[86] In these instances, the Chinese state not only views innovation as a solution to social ills and a means to nation building, but explicitly signals which types of innovation are permissible and desirable.

Geopolitical Platform Ecosystems

In late 2019, Alibaba filed for a secondary listing on the Hong Kong Stock Exchange (HKSE). This listing is meant to help reduce Alibaba's reliance on the US stock

TABLE 3. Common Institutional Investors in Chinese Platforms vs. GAFAM

INSTITUTIONAL INVESTOR	INVESTMENT IN CHINESE PLATFORMS	INVESTMENT IN GAFAM
SoftBank (Japan)	Alibaba	
Orbis Investment (South Africa)	NetEase, Sohu	
Baillie Gifford (UK)	Baidu, Tencent, Alibaba	Facebook, Microsoft, Alphabet, Amazon
T. Rowe Price (US)	Baidu, Sina	Facebook, Microsoft, Alphabet, Amazon
Schroder Investment Management (UK)	Sina	Facebook, Microsoft
BlackRock (US)	Sina, Alibaba, Tencent, Baidu	Facebook, Microsoft, Alphabet, Apple, Amazon
Macquaire Group (Australia)	Sohu	Facebook
Renaissance Technology (US)	Sohu	Facebook
JPMorgan Chase (US)	Tencent	Facebook, Microsoft, Alphabet
Hillhouse Capital (China)	iQiyi, Alibaba, JD, Sohu	Facebook, Apple, Amazon
Sequoia Funds (US)	Pinduoduo, JD, Sina, iQiyi, Alibaba	Alphabet, Facebook, Amazon
Lazard Asset Management (US)	Baidu	Facebook, Microsoft, Alphabet, Amazon, Apple
Vanguard Group (US)	Alibaba, Baidu	Apple, Microsoft, Alphabet, Amazon, Facebook
State Street (US)	Baidu	Apple, Microsoft, Amazon, Alphabet

market to access capital, as well as to ensure continuity in trading its stock in lieu of the worsening of US-China trade relationships. This decision proved to be prescient. In May 2020, the US Senate, with rare bipartisan support, approved legislation that forces Chinese companies to be more transparent in their financial reporting or face delisting from US stock exchanges.[87] Shortly after, China-based online gaming company NetEase and e-commerce platform JD pursued secondary listings on the HKSE.

The financial fallout of foreign laws specifically targeting Chinese platform companies could be significant as it would constrain their ability to raise capital. That said, on a financial level, the political economy of Chinese digital platforms is deeply integrated with global networks of investors, management, and capital.[88] Table 3 shows the degree to which both US-based and China-based platform companies are financed by similar groups of institutional investors. Not only are

Chinese and US platform companies owned by similar institutional investors, as Lee notes, these institutional investors, in turn, also own each other.[89] For example, T. Rowe Price is owned by Vanguard, BlackRock, and State Street, whereas BlackRock is owned by Vanguard and State Street. These complex and deeply interlocking relationships not only reinforce financial hegemony by institutionalizing power through ownership and reinforcing an elite managerial class,[90] they also showcase the interconnectedness of Chinese platform companies with global finance networks. Such political economy arrangements challenge the multipolarity of platform innovation, because they deepen and expand US–China alliances as well as the reach of capitalism and financialization.

The capitalist characteristics of Chinese platform companies position them as both collaborators and competitors with their US counterparts. Next to financial alliances there is infrastructural integration across platform ecosystems: WeChat and TikTok can be downloaded in global app stores, and citizens across North America and Europe are keen to order goods straight from Alibaba's e-commerce platform in China. The level of state control over the BAT platforms may be unchallenged and virtually unmatched, which sets China-based companies apart from the majority of their counterparts. At the same time, the integration of financial markets and "interplatform" infrastructures complicates the national identities of China-based tech companies.[91] It becomes increasingly difficult to pinpoint a clear association between their domestic origins and corporate behaviors.[92]

"If We Don't, China Will"

The pursuit of profit has increased competition and consolidation among Chinese and US digital markets. Fueled by two diametrically opposed political systems, shared concerns about national data sovereignty, and a competitive playing field, a geopolitical clash between both platform ecosystems seems all but inevitable. In September 2020, citing threats to national security, the US Department of Commerce banned WeChat and TikTok and ordered TikTok to delete all user data generated in the United States and further divestiture of its US operation.

With the world's largest internet economies, US and Chinese internet companies share the goal of devouring competition and expanding their global dominance. While as of yet, China-based internet companies lag behind in depth and breadth of their global offerings compared to their US counterparts, they are increasingly active as investors in both start-ups and incumbent enterprises.[93] Acquisition of start-up companies has always been Google's central strategy to increase market share beyond its primary business divisions (i.e., search and advertising), as evidenced by acquisitions of Keyhole (which later became Google Earth) in 2004, Android (2005), YouTube (2006), DoubleClick (2007), and many others.[94] As Google restructured to become a subsidiary of Alphabet, its financialization strategies became even more apparent by way of the establishment of three investing funds: GV (formerly Google Ventures), CapitalG, and Gradient Ventures. In 2017, these funds closed 103 deals, making Alphabet the most prolific corporate investor of the year. Then again, in the same year Tencent Holdings trailed Alphabet's shopping spree only slightly with 72 deals.[95]

Despite geopolitical tensions, US and Chinese funds have co-invested in a number of e- businesses focusing particularly on emerging Southeast Asian markets. In 2016, Google launched a multiyear e-Conomy SEA project together with Singapore sovereignty fund Temasek. Its goals are to make inroads into the region's blooming internet economy by investing in online travel, (digital) media, ride hailing, and e-commerce.[96] Under the auspices of the CCP's policy of "going out," Chinese digital platforms and investors started to gradually match efforts similar to the e-Conomy project. As a result, two axes of platform power mixing US/Chinese

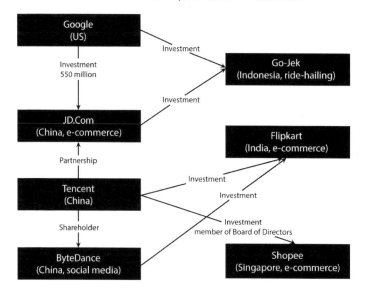

FIGURE 1. Chinese and US platforms in Southeast Asia

companies have emerged in the Southeast Asian market: Google-Tencent-JD.com and Amazon-Alibaba.

Through interlocking investments, shareholding agreements, and cross-appointments of board directors, Alphabet is partnered with JD.com, Tencent, and ByteDance to compete against Amazon. A Tencent board member is cross-appointed on the board of the Singaporean e-commerce company Shopee. These ties go beyond the financial level as they include deep infrastructural integrations. For example, Alibaba's Cloud services host Tokopedia and Lazada e-commerce services. Through investments in Chinese platform companies, Alphabet is able to take advantage of their cultural proximity to Southeast Asian markets and indirectly compete against its rival Amazon.

Just as Alphabet and Amazon both compete and cooperate, so do US platforms oscillate between institutional integration and clamoring for state support. The US-China fragmentation manifests itself through strategic and political mobilization of discourses around fundamental values, national sovereignty, and security.[97] US executives have a grab bag of discursive tools at their disposal to scare lawmakers into drafting "America First"–inspired legislation. There is the "we have to be big in order to beat China" trope to justify the growing market dominance of Amazon and Google. Pointing to China's ability to puncture holes in the United States' global data

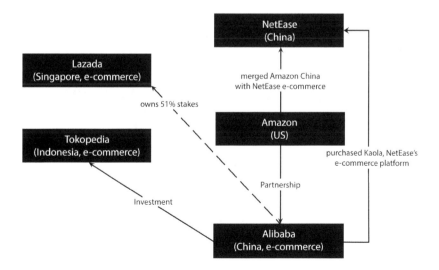

and market hegemony, Google engineering director Hartmus Neven stated: "We are indeed most worried (about) an unknown competitor out of China to beat us in the race to (such a) machine because China as a society just has the ability to steer enormous resources in the directions that are deemed strategically important."[98] This rhetorical approach falls into the "if we don't, China will" frame. In an attempt to fence off domestic regulatory scrutiny, Facebook's expansion into fintech and its investments in cryptocurrency venture Libra used this frame to great effect. In a 2019 hearing before the US House of Representatives, Facebook executive David Marcus argued: "I believe that if America does not lead innovation in the digital currency and payments areas, others will. If we fail to act, we could soon see a digital currency controlled by others whose values are dramatically different."[99]

Together with peddling the "threat of China" frame, reviewing Chinese investments, and the ongoing delisting and homecoming of Chinese companies from US stock exchanges, China mania has turned into China phobia. However, our chapter recognizes the multifaceted operationalization of platform power, which involves taking stock of platform histories, geographies, interlocking relationships, networks, and a complex global political economy.[100] This perspective is an important first step to get out of the binary thinking when considering China's rise as a global digital power and how it competes globally, particularly with the United States.[101] Our analysis shows that even though Chinese platforms harness the world's largest domestic user base, its global reach is still relatively limited. Indigenous innovation does not seem to be as globally exportable as the platforms and apps coming out of Silicon Valley.[102] As we noted, fueled by the COVID 19 crisis, platform capitalism "with Chinese characteristics" has started to face serious US political headwinds. Meanwhile, although the Chinese state closely streamlines its policy and developmental goals with the business expansion of Chinese platforms, it does not mean that they are commercially viable, nor that the platforms are always inherently acting as state proxies. Conversely, apart from innovative technology and a sizable domestic market, it is the unprecedented financial and infrastructural power of Chinese platforms that propels their ecosystems forward. The ability to attract finance capital or subsidize loss-making, long-term infrastructural investments with profit-making businesses can crowd out market competition. Thus, capital transcends national boundaries and brings US and Chinese platforms together as strange bedfellows to collectively devour emerging markets outside their home bases.

NOTES

1. See https://www.judiciary.senate.gov/meetings/how-corporations-and-big-tech-leave-our-data-exposed-to-criminals-china-and-other-bad-actors, and https://www.judiciary.senate.gov/meetings/dangerous-partners-big-tech-and-beijing.
2. Drew Harwell and Tony Romm, "Inside TikTok: A Culture Clash Where U.S. Views about Censorship Often Were Overridden by the Chinese Bosses," *Washington Post*, November 5, 2019.
3. See https://www.congress.gov/bill/116th-congress/senate-bill/3455.
4. Dal Yong Jin, Digital Platforms, Imperialism and Political Culture (New York: Routledge, 2015).
5. José Van Dijck, David Nieborg, and Thomas Poell, "Reframing Platform Power," *Internet Policy Review* 8, no. 2 (2019). For an in-depth critique of the challenge of anchoring public values in a "platform society" see José Van Dijck, Thomas Poell, and Martijn De Waal, *The Platform Society: Public Values in a Connective World* (Oxford: Oxford University Press, 2018).
6. "Topic: TikTok," Statista, 2020, https://www.statista.com/topics/6077/tiktok/.
7. Van Dijck, Nieborg, and Poell, "Reframing Platform Power."
8. Van Dijck, Poell, and De Waal, *The Platform Society*.
9. Junyi Lv and David Craig, "Firewalls and Walled Gardens: The Interplatformization of China's Wang Hong Industry," in *Engaging Social Media in China: Platforms, Publics, and Production*, ed. Guobin Yang and Wei Wang (East Lansing: Michigan State University Press, 2021).
10. Van Dijck, Nieborg, and Poell, "Reframing Platform Power."
11. David B. Nieborg and Anne Helmond, "The Political Economy of Facebook's Platformization in the Mobile Ecosystem: Facebook Messenger as a Platform Instance," *Media, Culture & Society* 41, no. 2 (2019): 196–218. Jean-Christophe Plantin, Carl Lagoze, Paul N. Edwards, and Christian Sandvig, "Infrastructure Studies Meet Platform Studies in the Age of Google and Facebook," *New Media & Society* 20, no. 1 (2018): 293–310.
12. Van Dijck, Nieborg, and Poell, "Reframing Platform Power."
13. Nieborg and Helmond, "The Political Economy of Facebook's Platformization in the Mobile Ecosystem."
14. Lianrui Jia and Dwayne Winseck, "The Political Economy of Chinese Internet Companies: Financialization, Concentration, and Capitalization," *International Communication Gazette* 80, no. 1 (2019): 30–59.
15. Zhengrong Hu and Deqiang Ji, "Ambiguities in Communicating with the World: The 'Going-out' Policy of China's Media and Its Multilayered Contexts," *Chinese Journal of*

Communication 5, no. 1 (2012): 32–37.
16. Julie Yujie Chen, "The Mirage and Politics of Participation in China's Platform Economy," *Javnost–The Public* 27, no. 2 (2020): 154–70.
17. Susan Leong, "Prophets of Mass Innovation: The Gospel According to BAT," *Media Industries Journal* 5 no. 1 (2018).
18. Julie Yujie Chen and Jack Linchuan Qiu, "Digital Utility: Datafication, Regulation, Labor, and DiDi's Platformization of Urban Transport in China," *Chinese Journal of Communication* 12, no. 3 (2019): 274–89.
19. Lin Zhang, "When Platform Capitalism Meets Petty Capitalism in China: Alibaba and an Integrated Approach to Platformization," *International Journal of Communication* 14 (2020): 118.
20. Jean-Christophe Plantin and Gabriele de Seta, "WeChat as Infrastructure: The Techno-Nationalist Shaping of Chinese Digital Platforms," *Chinese Journal of Communication* 12, no. 3 (2019): 257–73.
21. See Lin and de Kloet in this volume.
22. Xu Chen, D. B. V. Kaye, and J. Zeng, "#PositiveEnergy Douyin: Constructing 'Playful Patriotism' in a Chinese Short-Video Application," *Chinese Journal of Communication* 14, no. 1 (2020): 97–117.
23. Hong Shen, "Building a Digital Silk Road? Situating the Internet in China's Belt and Road Initiative," *International Journal of Communication* 12 (2018): 2683–701.
24. Hong Shen, "Building a Digital Silk Road."
25. Maximiliano Facundo Vila Seoane, "Alibaba's Discourse for the Digital Silk Road: The Electronic World Trade Platform and 'Inclusive Globalization,'" *Chinese Journal of Communication* 13, no. 1 (2020): 68–83.
26. Michael Keane and Haiqing Yu, "A Digital Empire in the Making: China's Outbound Digital Platforms," *International Journal of Communication* 13 (2019): 4624–41.
27. Qi Yin and Xiaoxia Li, "Exploring the Roles of Government Involvement and Institutional Environments in the Internationalization of Chinese Internet Companies," *Chinese Journal of Communication* 13, no. 1 (2020): 47–67.
28. Angelica Mari, "Baidu Ends Brazil Operations," *ZDNet*, July 17, 2018, https://www.zdnet.com/article/baidu-ends-brazil-operations/.
29. Mari, "Baidu Ends Brazil Operations."
30. Asa Fitch, "Chinese Tech Giant Baidu Steps Back from Middle East," *Wall Street Journal*, October 26, 2017, sec. Business, https://www.wsj.com/articles/chinese-tech-giant-baidu-steps-back-from-middle-east-1509039191.
31. Sarah Logan, Smith Graeme, and Molloy Brenda, "Chinese Tech Abroad: Baidu in

Thailand," Internet Policy Observatory, Center for Global Communication Studies, University of Pennsylvania (2018). Even though Baidu has not been a particularly successful case, other companies and products, such as Tencent and Alibaba, are better off, which further suggests globalization is an uneven and rocky process.

32. Leong, "Prophets of Mass Innovation."
33. Loretta Chao, "Baidu Brushes Up on Its Arabic, Thai," *WSJ* (blog), September 15, 2011, https://blogs.wsj.com/chinarealtime/2011/09/15/baidu-brushes-up-on-its-arabic-thai/.
34. Lianrui Jia, "Unpacking China's Social Credit System: Informatization, Regulatory Framework, and Market Dynamics," *Canadian Journal of Communication* 45, no. 1 (2020): 113–27.
35. Stanislav Budnitsky and Lianrui Jia, "Branding Internet Sovereignty: Digital Media and the Chinese–Russian Cyberalliance," *European Journal of Cultural Studies* 21, no. 5 (2018): 594–613.
36. Lin Zhang, "When Platform Capitalism Meets Petty Capitalism in China."
37. Plantin and de Seta, "WeChat as Infrastructure." Julian Gruin, "Financializing Authoritarian Capitalism: Chinese Fintech and the Institutional Foundations of Algorithmic Governance," *Finance and Society* 5, no. 2 (2019): 84–104. Jing Wang, "'The Party Must Strengthen Its Leadership in Finance!': Digital Technologies and Financial Governance in China's Fintech Development," *China Quarterly* (2020): 1–20, doi.org/10.1017/S0305741020000879.
38. Van Dijck, Nieborg, and Poell, "Reframing Platform Power." Hong Shen et al., "'I Can't Even Buy Apples If I Don't Use Mobile Pay?': When Mobile Payments Become Infrastructural in China," *Proc. ACM Human Computer Interaction* 4, no. CSCW2 (2020): article 170.
39. Celia Chen, "China's Mobile Payments to See Rebound as Offline Vendors Reopen after Coronavirus Lockdowns," *South China Morning Post,* April 2, 2020.
40. John Detrixhe, "China's Ant Financial Raised Almost as Much Money as All US and European Fintech Firms Combined," *Quartz,* January 30, 2019, https://qz.com/1537638/ant-financial-raised-almost-as-much-money-in-2018-as-all-fintechs-in-us-and-europe/. Divested by Jack Ma from Alibaba in 2011, Ant Financial is partially owned by Alibaba (33%) with a market capitalization estimated at $200 billion in 2020, roughly the market valuation of Goldman Sachs and Morgan Stanley combined, surpassing those of many traditional financial institutions.
41. Bien Perez, "Alipay Expands Alliance with US Payment Processor First Data," *South China Morning Post,* May 29, 2017.
42. Gopal Sharma, "Nepal Says Bans WeChat Pay, Alipay," *Reuters,* May 22, 2019, https://

43. www.reuters.com/article/us-china-nepal-digitalpayments/nepal-says-bans-wechat-pay-alipay-idUSKCN1SS19N/.
43. "China's Alipay, WeChat Pay Permitted to Provide Service in Nepal," *Xinhua*, March 1, 2020, http://www.xinhuanet.com/english/2020-03/01/c_138830881.htm.
44. Zen Soo, "TechFin: Jack Ma Coins Term to Set Alipay's Goal to Give Emerging Markets Access to Capital," *South China Morning Post*, December 2, 2016, https://www.scmp.com/tech/article/2051249/techfin-jack-ma-coins-term-set-alipays-goal-give-emerging-markets-access.
45. Dirk Andreas Zetzsche, Ross P. Buckley, Douglas W. Arner, and Janos Nathan Barberis, "From FinTech to TechFin: The Regulatory Challenges of Data-Driven Finance," *New York University Journal of Law and Business* 14, no. 2 (2018): 393–446.
46. Jing Wang and Mai Anh Doan, "The Ant Empire: Fintech Media and Corporate Convergence within and beyond Alibaba," *Political Economy of Communication* 6, no. 2 (2018), 25–37.
47. However, as of 2019, newly introduced regulation and market competition by both traditional banks and digital financial services made YuEBao decide to no longer offer such high annual rates. Zhang Qianqian, "Yuebao Annualized Rate Broke 2% for the First Time," *Xinhuanet*, April 9, 2020, http://www.xinhuanet.com/fortune/2020-04/09/c_1125833322.htm.
48. Stella Yifan Xie, "More Than a Third of China Is Now Invested in One Giant Mutual Fund," *Wall Street Journal*, March 27, 2019, https://www.wsj.com/articles/more-than-a-third-of-china-is-now-invested-in-one-giant-mutual-fund-11553682785.
49. Yuzhe Zhang and Timmy Shen, "New Credit Bureau Finds Good Data Is Hard to Come By," *Caixing Global*, May 22, 2019, https://ww.caixinglobal.com/2019-05-22/new-credit-bureau-finds-good-data-is-hard-to-come-by-101418792.html.
50. Patrick Shaou-Whea Dodge, "Communication Convergence and "the Core" for a New Era," in *Communication Convergence in Contemporary China: International Perspectives on Politics, Platforms, and Participation*, ed. Patrick Shaou-Whea Dodge (East Lansing: Michigan State University Press, 2021), ix–xxxii.
51. Julian Gruin, "Financializing Authoritarian Capitalism: Chinese Fintech and the Institutional Foundations of Algorithmic Governance," *Finance and Society* 5, no. 2 (2019): 84–104.
52. Jia, "Unpacking China's Social Credit System." Fan Liang, Vishnupriya Das, Nadiya Kostyuk, and Muzammil M. Hussain, "Constructing a Data-Driven Society: China's Social Credit System as a State Surveillance Infrastructure," *Policy & Internet* 10, no. 4 (2018): 415–53.

53. Liang et al., "Constructing a Data-Driven Society."
54. Liang et al., "Constructing a Data-Driven Society."
55. Ryan McMorrow, "Alibaba Pledges to Spend $28bn on Cloud Computing," *Financial Times*, April 20, 2020, https://www.ft.com.
56. Caixiong Zhang, "Tencent Inks Pact for Data Center Efforts," *China Daily*, December 26, 2018, https://global.chinadaily.com.cn/a/201812/26/WS5c22cd91a310d91214050d82.html.
57. Alibaba Cloud has stepped up its globalization effort in 2015 and Tencent Cloud in 2014. Kai Jia, Martin Kenney, and John Zysman, "Global Competitors? Mapping the Internationalization Strategies of Chinese Digital Platform Firms," in *International Business in the Information and Digital Age*, ed. Rob Van Tulder, Alain Verbeke, and Lucia Piscitello (Bingley, UK: Emerald Publishing, 2019), 187–216.
58. McMorrow, "Alibaba Pledges to Spend $28bn on Cloud Computing."
59. US company Oracle also has a role in this; see Mara Hvistendahl, "Exclusive: How Oracle Sells Repression in China," *The Intercept*, February 18, 2021, https://theintercept.com/2021/02/18/oracle-china-police-surveillance/. Liza Lin and Josh Chin, "China's Tech Giants Have a Second Job: Helping Beijing Spy on Its People," *Wall Street Journal*, November 30, 2017, https://www.wsj.com/articles/chinas-tech-giants-have-a-second-job-helping-the-government-see-everything-1512056284.
60. "City Brain Now in 23 Cities in Asia," *Alibaba Clouder*, October 28, 2019, https://www.alibabacloud.com/blog/city-brain-now-in-23-cities-in-asia_595479.
61. WeChat ID contains personal information, domicile, personal identification number, and a photo. Celia Chen, "China's Social Security System Turns to WeChat for Electronic ID," *South China Morning Post*, January 5, 2018, https://www.scmp.com/tech/china-tech/article/2127010/chinas-social-security-system-turns-wechat-electronic-id.
62. The onset of COVID-19 expedited the achievement of WeChat's and Alipay's status as digital utility, as individuals must obtain and display a green health code to access public services, such as subways, highways, public transportation, or hospitals.
63. Anqi Fan, "WeChat Messages to Be Used as Evidence in Court," *China.org*, Feburary 5, 2015, http://www.china.org.cn/china/2015-02/05/content_34742048.htm. Celia Chen, "Chinese Court Moves to Streamline Admission of Wechat and QQ Chats in Civil Disputes," *South China Morning Post*, July 20, 2018, https://www.scmp.com/tech/china-tech/article/2156181/chinese-court-moves-streamline-admission-wechat-and-qq-chats-civil.
64. Chen, "Chinese Court Moves to Streamline."
65. The Supreme People's Court, *Chinese Courts and Internet Judiciary* (Beijing: People's

66. Van Dijck, Poell, and De Waal, *The Platform Society*.
67. "人工智能创新平台再添生力军 (AI Open Innovation Platform Added New Members)," *People's Daily*, October 15, 2019, http://www.xinhuanet.com/info/2019-10/15/c_138472595.htm.
68. For more discussions on the history of funding and ties to global capital for Chinese internet companies, see Jia, "Going Public and Going Global."
69. The five members are Jack Ma, Joe Tsai, Daniel Zhang, Lucy Peng, and Eric Jing.
70. Bingqing Xia and Christian Fuchs, *The Financialisation of Digital Capitalism in China* (London: Westminster Institute for Advanced Studies, 2016).
71. Jia and Winseck, "The Political Economy of Chinese Internet Companies."
72. Xia and Fuchs, "The Financialisation of Digital Capitalism in China."
73. Alibaba Group Holding Limited, *Yahoo! Finance*, https://finance.yahoo.com/quote/BABA/. Tencent Holdings, *Yahoo! Finance*, https://finance.yahoo.com/quote/TCEHY?p=TCEHY.
74. Amazon, *Yahoo! Finance*, https://finance.yahoo.com/quote/AMZN?p=AMZN. Alphabet, *Yahoo! Finance*, https://finance.yahoo.com/quote/GOOG/?p=GOOG. Facebook, *Yahoo! Finance*, https://finance.yahoo.com/quote/FB?p=FB.
75. Don Weinland, "Alibaba and Tencent Join State-Owned Groups in $11.7bn China Unicom Investment," *Financial Times*, August 16, 2017.
76. "CTC and Alibaba Team Up for Cloud Services," *China Daily*, August 1, 2018, https://europe.chinadaily.com.cn/a/201808/01/WS5b60faa1a31031a351e91540.html.
77. See also the commercial logics of state-sponsored platformization discussed in Guobin Yang, "Introduction: Social Media and State-Sponsored Platformization," in *Engaging Social Media in China: Platforms, Publics and Production*, ed. Guobin Yang and Wei Wang (East Lansing: Michigan State University Press, 2021), xx.
78. Min Tang, "From 'Bringing-in' to 'Going-out': Transnationalizing China's Internet Capital through State Policies," *Chinese Journal of Communication* 13, no. 1 (2020): 27–46.
79. 2018 Annual Report, Tencent (2019).
80. David B. Nieborg, C. J. Young, and D. J. Joseph, "App Imperialism: The Political Economy of the Canadian App Store," *Social Media + Society* 6, no. 2 (2020): 1–11.
81. Costas Lapavitsas, "The Financialization of Capitalism: Profiting without Producing," *City* 16, no. 6 (2013): 792–805.
82. Johannes Petry, "Financialization with Chinese Characteristics? Exchanges, Control and Capital Markets in Authoritarian Capitalism," *Economy and Society* 49, no. 2

(2020): 213–38.
83. Chen, "The Mirage and Politics of Participation in China's Platform Economy."
84. See Zhang Lin, "Assembling Alibaba: The Infrastructuralization of Digital Platforms in China," in *Engaging Social Media in China: Platforms, Publics, and Production*, ed. Guobin Yang and Wei Wang (East Lansing: Michigan State University Press, 2021).
85. In 2020, the Chinese government stepped up its efforts on platform monopoly.
86. See for a discussion on Weibo: Lianrui Jia and Xiaofei Han, "Tracing Weibo (2009–2019): The Commercial Dissolution of Public Communication and Changing Politics," *Internet Histories* 4, no. 3 (2020): 304–32, https://doi.org/10.1080/24701475.2020.1769894; People's Daily, "人民日报评社区团购：别只惦记几捆白菜！科技创新更令人心潮澎湃!" *Sohu*, December 14, 2020, https://www.sohu.com/a/438055555_650579.
87. See https://www.congress.gov/bill/116th-congress/senate-bill/945.
88. Jia, "Going Public and Going Global." Jia and Winseck, "The Political Economy of Chinese Internet Companies."
89. Micky Lee, *Alphabet: The Becoming of Google* (London: Routledge, 2019).
90. Gerard Dumenil and Dominique Lévy, *Capital Resurgent: Roots of the Neoliberal Revolution* (Cambridge, MA: Harvard University Press, 2014).
91. Junyi Lv and Craig, "Firewalls and Walled Gardens.'"
92. Dwayne Winseck, "The Geopolitical Economy of the Global Internet Infrastructure," *Journal of Information Policy* 7 (2017): 228–67.
93. Yin and Li, "Exploring the Roles of Government Involvement and Institutional Environments in the Internationalization of Chinese Internet Companies"; Jia and Winseck, "The Political Economy of Chinese Internet Companies"; Tang, "From 'Bringing-in' to 'Going-Out.'"
94. Lee, *Alphabet*.
95. Jason Rowley, "A Peek inside Alphabet's Investing Universe," *TechCrunch*, February 17, 2018, https://techcrunch.com/2018/02/17/a-peek-inside-alphabets-investing-universe/.
96. "e-Conomy SEA 2018," Google Temasek, 2018, https://www.thinkwithgoogle.com/intl/en-apac/tools-resources/research-studies/e-conomy-sea-2018-southeast-asias-internet-economy-hits-inflection-point/.
97. Dodge, "Introduction: Communication Convergence and 'the Core' for a New Era."
98. Sintia Radu, "New Warnings over China's Efforts in Quantum Computing," *USNews*, January 31, 2020, https://www.usnews.com/news/best-countries/articles/2020-01-31/google-quantum-chief-warns-china-can-quickly-develop-supercomputers.
99. David Marcus, "Hearing before the United States House of Representatives Committee on Financial Services: Testimony of David Marcus," *United States House Committee on*

Financial Services, July 17, 2019, https://www.congress.gov/116/meeting/house/109821/witnesses/HHRG-116-BA00-Wstate-MarcusD-20190717.pdf.

100. Marc Steinberg, *The Platform Economy: How Japan Transformed the Consumer Internet* (Minneapolis: University of Minnesota Press, 2019).
101. Wilfred Yang and Ramon Lobato, "Chinese Video Streaming Services in the Context of Global Platform Studies," *Chinese Journal of Communication* 12, no. 3 (2019): 356–71.
102. However, for device manufacturers such as Transsion, it is a different story. See Lu in this volume, and a discussion on maker culture in China by Mutibwa and Xia in this volume.

Neoliberal Business-as-Usual or Post-Surveillance Capitalism with European Characteristics?

The EU's General Data Protection Regulation in a Multipolar Internet

Angela Daly

Since the 2008–2009 Global Financial Crisis the tenets of neoclassical economic theory and its application through neoliberalism have come into question. Here, I understand neoliberal capitalism as "a programme of resolving problems of, and developing, human society by means of competitive markets."[1] In particular, there has been a move away from the idea of "light touch" non-interventionalist regulatory approaches in Western economies, including the European Union (EU) and increasingly also the United States—a process that may be accelerated by the current COVID-19 pandemic and its fallout.[2] In addition, the Snowden revelations of US-driven global digital surveillance prompted law and policy changes in various jurisdictions, including the EU, to counter privacy infringements by both state and corporate actors. Such changes prompt an attempt at legal innovation in the wake of technological communication innovation and the emergence of new business models revolving around the exploitation of user data. Moreover, these changes take place against a backdrop of an increasingly multipolar world, including in matters of internet governance, where the United States' hegemony is weakening and the EU and BRICS, notably China with its vast and increasingly globalized internet and digital technologies industry, emerge as powerful actors, including in the internet sphere. Indeed, the aftermath of COVID-19 may well see a strengthened position internationally for

China as one half of a bipolar order with the United States.³ The Chinese internet industry brings problems of its own with a surveillance-industrial complex overseen by the Chinese government in its Social Credit System—one of many ways in which Chinese internet companies provide surveillance infrastructure to the authoritarian-capitalist Chinese state.⁴

One major event with international impact for the EU is the updating of its data protection legislation, with the introduction of the General Data Protection Regulation (GDPR) in 2016. The GDPR is widely believed to be the international "gold standard" in privacy and data protection standards, holding actors in the digital economy to account for their (ab)use of personal data and thereby empowering users by seemingly placing some limitations on unfettered digital capitalism and surveillance.⁵ Can we understand the GDPR developments as an attempt by the EU to forge a different political economy of digital technology to neoliberal capitalism from the United States and state-led authoritarian capitalism in China, whereby large companies from both locations are constrained in their actions and EU citizens' rights upheld? Or can these developments be seen through another lens, that of varieties of capitalism, whereby the European Union, in the context of growing geopolitical multipolarity, is leveraging its large and attractive market to become a "regulatory superpower" producing legal innovations after having lost the battle to become a communication innovation superpower to the United States and China?⁶

This question evokes the notion of the "Brussels Effect": the EU's "unilateral power to regulate global markets" by "externaliz[ing] its laws and regulations outside its borders through market mechanisms."⁷ While this Brussels Effect was originally more incidental to the EU's internal goals (e.g., establishing the EU Single Market), in recent years, the EU is increasingly conscious of the extraterritorial reach, whether de facto or de jure, of its regulatory activities. In embracing this extraterritorial reach, the EU may be pursuing objectives including securing the competitiveness of EU industry, obtaining "greater legitimacy for its rules," and the expansion of "EU's soft power and validating its regulatory agenda, both at home and abroad."⁸

From here, I provide background to the GDPR case study. First, I explain EU internet regulation, and more generally the EU's interactions with neoliberalism in its regulatory and governance activities, before moving on to a consideration of the Snowden revelations and their impact on internet regulation, innovation, and privacy. I present the emerging geopolitics of internet governance before I examine and analyze the GDPR itself.

The GDPR exposes the nature of contemporary EU internet regulation as a contested and hybrid site, containing both capitalist impulses and overtures to protect individuals' privacy and data. Yet protecting privacy and data is diminished to the extent that the GDPR facilitates significant data gathering and use, and seems to have the effect in practice of consolidating US tech giant Google's corporate power. Thus, the GDPR seems more an example of EU regulatory capitalism, "constraining and encouraging the spread of neoliberal norms,"[9] rather than a legally innovative step on the path towards digital postcapitalism.[10] Nevertheless, the extraterritorial reach of the GDPR makes it an example of the Brussels Effect and contributes to shoring up the EU's regulatory power in a multipolar internet by proactively setting de facto standards in data protection. In this way, the EU "remains relevant as a global economic power" through its legal innovation, rather than through its technological innovation like the United States and China.[11] Despite the fact that this technological innovation may come from US and Chinese Big Tech companies, the GDPR may also trigger more privacy-friendly innovation through its human rights and social goals aspects. Yet this privacy-friendly innovation triggered by the GDPR may not undermine the power of US and Chinese Big Tech firms and may instead give rise to a more limited or moderate form of surveillance capitalism practiced in the EU.

EU Internet Regulation and Neoliberalism

The internet's historical origins are in the United States, although collaborations commenced with researchers in other parts of the world, including Europe, early in its development. Regulatory and policy activity only properly commenced in the 1990s, when the internet started to become publicly available. During this decade there was limited EU regulatory and policy activity in this sphere: according to Feeley, the EU's objective was "to protect society and create an equitable internet environment."[12] Avoiding fragmentation of the EU's Single Market has been, at least rhetorically, a common theme justifying the EU's forays into internet regulation and policymaking from the 1990s onwards.[13]

Since the late 1990s, industry self-regulation emerged as another tool of EU internet regulation and policymaking. This may have mirrored the (at the time) hegemonic and capitalist US approach and may also have aimed at facilitating EU industry's commercial entry into the emerging internet markets.[14] Christou and

Simpson point to the internet's US origins in a self-regulatory environment and the prevailing discourse of neoliberal globalization as influencing EU internet regulation.[15] More recently, the EU has exhibited a public/private cooperation approach to global internet governance and regulation.[16] Marsden has characterized this as "co-regulation" whereby "the regulatory regime is made up of a complex interaction of general legislation and a self-regulatory body," with the state actor often delegating a large measure of responsibility for making and applying rules to the particular industry sector, as can be seen in content regulation and domain name governance.[17]

The EU pursues different objectives and approaches in its contemporary internet regulation, displaying, according to Halpin and Simpson, "a 'mixed mode' of governance combining, on the one hand, acceptance of the neo-liberal model of self-regulation with, on the other, a distinctly more interventionist 'hands on' policy with specific commercial and social goals in mind."[18]

Halpin and Simpson point to the neoliberal policy agenda dominating the EU's approach to the internet but not completely subsuming it, as the EU's approach also embraces social goals. Key features of neoliberalism pertinent to telecommunications and internet regulation are the privatization of previously state-owned utilities, the liberalization of markets, and deregulation/regulatory forbearance.[19] Some implementations of EU internet regulation involve co-regulation between government and industry, and other implementations involve a self-regulatory approach by industry, which is more in line with neoliberal norms.[20]

In 2008–2009, a major financial crisis hit the global economy, which originated in the United States, with significant impact felt in the EU. It has been termed "a systemic crisis of neoliberal capitalism" since it was caused by features of the neoliberal system such as deregulation of business (especially finance) leading to growing inequality, financial speculation, and asset bubbles.[21] The 2008–2009 financial crisis challenged previously prevailing hegemonic thought, especially in Western countries, on neoliberal capitalism being the best or even a good political economic system. In particular, the crisis challenged privatized liberal markets and deregulation as desirable or optimal policy approaches to economic governance, opening up possibilities for other political economy theories to guide regulation and policy. Such possibilities encompass, for instance, regulation and policy that initiate a post-globalization trend by strengthening state power over surveillance-capitalist markets or that stimulate a "postcapitalist" orientation of technological development altogether. The development of information technologies may serve as a further impetus to moving beyond (neoliberal) capitalism to a postcapitalist world where digitization and automation could perform labor, diminishing the need for work and thereby undermining the tenets and assumptions of neoliberal capitalism.[22] However, postcapitalism will not inevitably flow from digitization and automation, and a number of intervening transformations in technological, social, and economic practices are needed for digital technologies to eventuate in such a future.[23]

In fact, in the intervening years the EU has not instigated a major change in policy direction, and instead imposed austerity policies—a neoliberal instrument.[24] Furthermore, many of the benefits of digital technologies have not been shared widely in accordance with a postcapitalist economy and instead have been captured by large companies and rich individuals in an increasingly "platform capitalist" economy.[25] This seems to be intensified by the COVID-19 pandemic as digital technology companies benefit from the increase in remote online activity and app-based delivery services.[26]

Despite the potential inherent in material conditions and technology for a different approach, the EU's internet regulation post-crisis represents a continuity with its previous approach identified by Halpin and Simpson, with elements of neoliberalism interspersed with other social and economic goals.[27] An example is the net neutrality regulation introduced in 2015, which, according to Robin-Olivier, "illustrates the entanglement of fundamental freedoms (freedom of speech, freedom of information and pluralism), economic rights and interests, and democratic values."[28]

Snowden Revelations, Online Surveillance, and the Social Credit System

The Snowden revelations in 2013 were a major event for internet governance internationally. Edward Snowden, a former US National Security Agency (NSA) employee and then contractor, revealed that the NSA and its partners in the "Five Eyes" (FVEY) surveillance partnership (Australia, Canada, New Zealand, and the United Kingdom) had engaged in secretive mass surveillance and data-gathering operations including through co-option of large US-based internet companies such as Google and Yahoo.[29] It remains unclear to what extent these companies voluntarily cooperated with the US authorities or were hacked. However, their large physical and virtual infrastructures provide key aspects of the internet experience for users worldwide. In doing so, these companies collect large amounts of data through their products and services. Indeed, for many companies, notably Google, this huge data collection is a key element of their business models.[30] Overall, this preexisting infrastructure proved highly attractive to the US NSA surveillance apparatus. At the time, due to the internet's history and origins in that country, the United States was a key conduit for global internet flows. Large amounts of data and information passed through US-based infrastructure, coupled with a global policy of "internet freedom" to promote US state power and ensure markets were open to its companies.[31]

This data gathering by US-originating multinational internet companies such as Google proved convenient to state security forces in the United States and partner countries. However, this corporate data gathering is motivated principally by economic rationales; the gathering, processing, and analysis of data is a key aspect of these companies' business models, especially when their services are "free" in the sense of bearing no monetary cost to internet users. Thus, surveillance of user activities is initially carried out for economic purposes, and constitutes what Zuboff has termed "surveillance capitalism."[32] Yet this user surveillance also proves highly convenient to nation-states, which encourages them to engage in an "invisible handshake" with these companies.[33] There has been limited regulation of these companies' practices, which reflects neoliberal "light touch" regulatory ideology cautioning against "interference" with private companies' activities and the convenience of these data stocks for government security apparatuses.[34]

The 2018 Cambridge Analytica scandal represents a further key development in this area. Cambridge Analytica and its associates had been harvesting individuals' data from their Facebook profiles since 2013 without their clear knowledge and consent. This data was fed into political campaign advertising microtargeted at individuals for Cambridge Analytica's clients. While infringements of data protection law were identified,[35] the damage is done inasmuch as the data has already been created, used, and analyzed—which may have influenced the outcome of major political events in the West such as the 2016 US presidential election and the UK's 2016 referendum on EU membership.[36]

Aho and Duffield characterize the EU's GDPR as a "reactive response" to this (US-driven) surveillance capitalism, which can be contrasted with China's "proactive response" in the form of its Social Credit System (SCS), whereby China "embraces" surveillance capitalist "logics for further state use."[37] The SCS operates by assigning credit scores to both companies and individuals within China, analyzed by algorithms "operationally managed by central government authority," allowing the state to incentivize "good" behaviors and punish "bad" behaviors through "an operationally managed system of tailored rewards and punishments."[38] In doing so, the SCS collects enormous amounts of data about individuals and companies, including from digital and internet companies, to facilitate the Chinese authoritarian-capitalist state's achievement of social and economic control.[39] The SCS comes against the backdrop of increasing surveillance and censorship of the Chinese internet since President Xi Jinping came to power in 2012, which is particularly acute in Xinjiang as part of the Chinese government's severe repression of the Uyghurs and other minority groups there.[40]

In the West, China, and elsewhere, internet users face problems in protecting their digital rights such as privacy and avoiding chilling effects on their free expression in the West,[41] or outright censorship in China.[42] Users' digital rights are at risk from this "invisible handshake" of large companies and governments—or in the case of China, a much more visible and overt handshake—whose interests lie in perpetuating the generation and use of such data. Due to the size of large internet companies, the crucial physical and virtual infrastructure they provide, and the data they collect, they are in very powerful positions to manipulate our communications in both sociopolitical and economic ways. In a neoliberal setting, these companies may face limited oversight by governments, which also benefit from the data that is gathered.[43] In China's authoritarian capitalist setting, there is also an incentive for more data to be gathered about users by companies and public agencies that serves the state's SCS objectives.[44]

Geopolitics of the Internet

The aforementioned Snowden revelations and a realization that the United States' central place in the internet infrastructure of the 2010s—including through private companies headquartered in its territory—facilitated its disproportionate dominance and ability to surveil and gather data, prompted reactions from other countries. Specific examples of the NSA and partners spying on heads of state heightened some countries' reactions. One such example is that of Brazil, whose president at the time, Dilma Rousseff, was victim to FVEY spying on her mobile phone.[45] The macro and micro fallout of the Snowden revelations in Brazil led to the Netmundial conference, which sought to decentralize internet governance from the US/Global North, and to establish the Marco Civil "Internet Bill of Rights" domestically.[46] Internal political events in Brazil disrupted these activities, whereby Rousseff was impeached, and the governments since then have not continued this work.[47]

While political events ultimately arrested the Brazilian trajectory, the move towards internet and digital sovereignty by nation-states, especially but not exclusively large non-Western geopolitical players, has continued apace.[48] An attempt to relocate internet governance from the (Western) multistakeholder model to the multilateral United Nations system, specifically within the International Telecommunications Union (ITU), occurred in 2012, led by China and Russia, yet it was successfully opposed by countries including the United States and EU member states.[49]

Russia and China have moved toward their own visions of internet and data sovereignty in the intervening years.[50] They have adopted various measures to "nationalize" their internets, which not only facilitate political control over the medium but may also stimulate their own internet economies.[51] This has particularly been the case for China. For some time, even preceding the Snowden revelations, China has been nurturing its own national internet political economy, through economic stimulation for equipment manufacturing, the creation of virtual products and services, and the (partial) exclusion of some US Big Tech firms, notably Google and Facebook, which did not agree to comply with Great Firewall and other censorship restrictions.[52]

China now has an internet ecosystem parallel to the West's, with its own tech giants "BAT" (Baidu, Alibaba, and Tencent) dominating it in a similar way to "GAFA" (Google, Apple, Facebook, Amazon) in the West/rest. The Chinese internet infrastructure also facilitates the creation of the SCS, itself an aforementioned "proactive response" to US surveillance capitalism.[53] With the rise of breakthrough Chinese apps such as TikTok attracting a global user-base, the Chinese government's Digital Belt and Road Initiative and Chinese hardware providers such as Huawei prompting a retreat to economic nationalism in the United States (continuing under the Biden administration)[54] and possibly Western Europe as well, the Chinese internet is globalizing.[55] As discussed by Jia and Nieborg earlier in this volume, the Chinese internet and its corporate players may challenge—to some extent at least—the US/Western political and economic dominance of the medium—although Chinese competition may not present a challenge to the capitalist underpinning of the internet's political economy.[56]

The EU itself occupies a complex position as an important geopolitical player in the form of the world's largest trading bloc, albeit one that is not always unified or effective politically.[57] It has an overall capitalist character, albeit one with more social democracy than the United States, but the balance struck differs depending on the specific member state.[58] Relations with the United States have been cordial in the post-WW2 period, heightened in the post-Soviet period where more countries came within the West's orbit, some of which became EU and NATO member states.[59] Some EU member states have had particularly close relations with the United States, including on data and surveillance matters. The UK (which has only just left the EU at the time of writing) is a close partner of the US surveillance authorities in the FVEY alliance, while other EU members have been in a second or third tier of partnerships.[60] However, individuals in the EU have also been targets of surveillance activities, including heads of state such as German Chancellor Angela Merkel.[61]

The fallout from the Snowden revelations in (continental) Europe did result in some discussions about Europeanizing internet infrastructures, such as building a European "backbone" (to ensure that communications from a European source to a European destination remain in European networks and do not have to be routed through the United States) and creating a "European cloud."[62] Despite attempts to stimulate the European tech economy, the EU has not seen the creation of true competitors to US-based GAFA such as in China.[63] Indeed the EU market remains fully open to these US companies, which contrasts with the situation in Russia and China, for instance. The Latin alphabet and shared cultural norms may also facilitate the presence of US companies in the EU market, more so than in the Chinese or Russian markets. Of the BRICS, Brazilian, Indian, and South African internet markets are also characterized by the presence/dominance of these large US-based companies—but also increasingly Chinese-based internet companies.[64]

With the development of artificial intelligence (AI) (digital technology that uses digital data and may be internet-connected) the EU is also widely viewed as trailing the world leaders of research and development, the United States and China.[65] Nevertheless, the EU has itself begun to use the language of "sovereignty" in digital policy.[66] A notable example is the 2020 European Commission Communication "Shaping Europe's Digital Future."[67] This asserts that the EU must "create the right conditions . . . to develop and deploy its own key capacities, thereby reducing our dependency on other parts of the globe for the most crucial technologies," while at the same time the EU must remain open to players from other parts of the world willing to abide by EU rules and standards.[68]

So, while the EU does not have its own dominant internet industry and instead is a market open to the United States and increasingly China and their companies, the EU seems to be taking on a leading role in regulating and governing internet companies operating in its territory, especially in the wake of the Snowden revelations. On issues of privacy and data protection, and on curbing the economic power of large internet companies, the EU has taken active steps to further what could be seen as progressive positions vis-à-vis the power of large internet companies—a possible "post-globalization" trend in nation-states being willing to regulate global companies in digital markets.[69] Accordingly, the GDPR forms a case study to test the question of whether the EU is adopting a post-surveillance capitalism approach through its regulatory activity, or whether it is "business as usual" for the EU's model of regulatory capitalism.

The EU's General Data Protection Regulation

The EU has emerged as the global leader on privacy and data protection standards, especially with the implementation of its General Data Protection Regulation (GDPR) in 2018. Since the coming into force of the EU's Charter of Fundamental Rights in 2009, privacy and data protection each have the status of fundamental "constitutional" right.

The situation in the EU contrasts to varying extents with other large geopolitical players. The United States notably has no comprehensive federal-level data protection legislation, nor a full constitutional right to privacy. India at the time of writing has no data protection law, although the Indian Parliament is considering a bill.[70] There is no explicit right to privacy in the Indian Constitution, but its Supreme Court did find an implicit right in a 2017 case.[71] China has an emerging yet hitherto piecemeal law of data protection.[72] It released comprehensive draft legislation on data protection in late 2020 that reflects some—but not all—aspects of the GDPR.[73] There is also significant skepticism about the efficacy of China's data protection framework, especially as regards government use of data and forms of data protection that may impede economic development.[74] Brazil has recently adopted its own General Data Protection Law (*Lei Geral de Proteção de Dados* or LGPD), modeled on the GDPR, due to come into force in January 2021.[75] In Russia there is the "traditional repertoire of legal protections for the confidentiality of private communications and 'private life'" along with data protection legislation, but recent developments in internet sovereignty and data localization laws are increasingly viewed by citizens as "threats to individual privacy."[76] The South African Constitution recognizes explicitly the right to privacy, and the jurisdiction also has data protection legislation that came into force in 2020 and also reflects aspects of the GDPR.[77] While these are differentiated approaches to data protection, the GDPR's influence over developments in the BRICS is clear, demonstrating its extraterritorial reach.

In the post–Snowden revelations era, EU privacy and data protection laws have curtailed the surveillance capitalist practices of companies such as Google and of nation-states. There is an important "trilogy" of cases from the Court of Justice of the EU (CJEU) that seem to be a response to surveillance capitalism and the Snowden revelations: the invalidation of the Data Retention Directive in *Digital Rights Ireland*; the right-to-be-forgotten case against Google, *Costeja*; and the *Schrems* case invalidating the EU-US safe harbor agreement for personal data transfers.[78] These cases predate the GDPR's coming into force and demonstrate the importance that the CJEU attaches to privacy and data protection rights against the practices of both nation-states and (US) corporations.

The GDPR itself is a major reform to EU data protection legislation, which was finalized in 2016 and came into force in 2018. It repeals the previous legislation, the Data Protection Directive from 1995, and aims to update the rules for the current digital age. The GDPR includes a number of principles for the processing of personal data; the rights of "data subjects"; rules and obligations for data processors and controllers; rules for the transfer of data to "third countries" (i.e., non-EU member states); provisions relating to the supervisory authorities in each EU member state, and their cooperation and interaction with each other; and provisions on procedural rules. Some of these provisions existed already in the Data Protection Directive; some, such as the right to erasure, were previously established in CJEU case law; and some, such as the right to data portability, are new introductions.

The GDPR has an extraterritorial reach contained in its Article 3 on Territorial Scope (and Recital 24), by which some entities located outside of the EU processing the data of EU residents may have to adhere to the provisions (at least according to EU law). This appears to function as an "anti-circumvention mechanism" for entities based outside of the EU handling EU data, so that they cannot escape the GDPR.[79] Furthermore, entities within the EU processing the data of non-EU citizens (and possibly non-EU residents) may also have to adhere to the law.

Analysis

The GDPR is widely considered as the leading data protection legislation globally, something of a "gold standard," as already mentioned. Many businesses have changed their practices since the GDPR came into force, whether or not they are based in the EU or exclusively handle EU residents' data: for some it might be cost-effective to apply GDPR standards to all data they handle.[80] Some non-EU jurisdictions have adopted or are considering GDPR-type protections in their own legislation, such as the Consumer Privacy Act 2018 in the US State of California[81] and Australia's Consumer Data Right,[82] along with the aforementioned example of Brazil—and to some extent China—modeling its own data protection legislation on the GDPR.

This "Brussels Effect" of the GDPR is not without controversy. There are no globally agreed standards of international law on data protection. Arora has critiqued the GDPR's implicit "privacy universalism" in elevating a Western/European conception of privacy and data protection to that of a global standard that may not be appropriate or useful in Global South contexts with different cultures of privacy.[83] Accordingly, even if well-intentioned, the GDPR as a de facto global standard "may inadvertently hinder the Global South's digital participation and become a neocolonial entity instead."[84]

The GDPR's rules for transferring personal data out of the EU (through "adequacy decisions" under GDPR Article 45 and other mechanisms), requiring appropriate safeguards for the data transferred, are problematic aspects of the EU's approach in an international context.[85] The EU's external trade policy reinforces this, whereby trade agreements concluded with non-EU "third" countries must be compatible with internal EU rules and policies, a stance that raises its own questions about the EU's compatibility with international trade law.[86] This extraterritorial reach of EU data protection standards "profoundly impacts suppliers of goods and services from outside the EU."[87] It also affects non-EU "third countries" themselves, including in their trade interactions and partnerships with other third countries. If such third countries agree to free cross-border data flows in another trade agreement, this may impact upon these third countries' interactions with the EU and might preclude them from obtaining an "adequacy finding" for their data protection laws.[88] These measures reinforce EU data protection norms' extraterritorial impact and are examples of their "Brussels Effect." This departs from a neoliberal approach to deregulation of personal data, but imposes the EU's view in the absence of a more multilateral global response to the issue, thereby reinforcing the EU's relevance as a global economic player in the context of technology governance via the legal innovation of the GDPR.

The GDPR's political economy aspects and the extent to which it may stimulate a GDPR-compliant EU digital market remain understudied aspects of the legislation. This is despite the fact that EU data protection law has long had a hybrid nature in protecting the human right to privacy but also facilitating trade in personal data within the EU Single Market—objectives contained within Article 1 of the GDPR.[89] Accordingly, the GDPR is not an inherently anti-capitalist piece of legislation,[90] and there are tensions within the GDPR's substance between these two objectives. However, there are also synergies between these objectives given that both reflect the Western/European liberal tradition in terms of both liberal humanism (human rights) and economic liberalism (trade in data).[91] While aspects of the GDPR give individuals some control over their data, there are still various deficiencies and ambiguities.[92] In many senses, the GDPR is ultimately permissive of various surveillance capitalist data-gathering and processing practices: it sets some bounds and restrictions on them but does not fundamentally alter the paradigm because of its objective of facilitating trade in personal data. The text of the GDPR itself is the product of a compromise between different interest groups including US Big Tech players, which heavily lobbied EU legislators and policymakers during its negotiation and formation.[93] Furthermore, the GDPR entered into force in a scenario where pervasive data gathering and analysis exist, often conducted by transnational companies based in the United States or China.[94]

Given the path dependency arising from this scenario, it would have been nothing short of revolutionary in all senses of the word if the GDPR had fundamentally disrupted this paradigm in both its substance and enforcement.[95] This evidences the GDPR as a rather belated intervention arriving after the fact of data-driven innovation, a key critical issue for innovation and technology governance acknowledged in the introduction to this volume. There are further aspects of the GDPR that suggest it does not represent a radical departure from the aforementioned previous EU internet regulation and governance approaches: Section 5 on Codes of Conduct and Certification evidences a co-regulatory approach whereby self-regulation by industry, particularly in the area of technical standardization, is combined with legal compliance overseen by public regulatory authorities.[96]

The GDPR's political economic effects are beginning to receive some attention in the literature. Peukert et al. have noted that post-GDPR some firms switched to using EU-based web technology providers.[97] However, Peukert et al. have also noted the increase in US tech giant Google's share in EU web technology markets since the GDPR came into force.[98] Geradin et al. reinforce the idea of Google's advantage from the GDPR, observing that the GDPR's implementation, rather than negatively impacting Google's dominance in the EU ad tech markets, has actually strengthened Google's position, as smaller rivals have found it difficult and costly to implement the new rules and operate in accordance with the framework.[99] Furthermore, Geradin et al. point to a lack of enforcement, particularly from the Irish data protection authority (under whose supervision Google falls in the EU) of the GDPR vis-à-vis Google's internal data practices, which has had a beneficial effect on Google's business. This may dampen the GDPR's effect in stimulating an EU digital market that could potentially then "export" its services to other parts of the world that have adopted similar laws following the GDPR's Brussels Effect. Ironically, rather than giving European companies a competitive advantage, the GDPR seems to be consolidating the power of US tech giant Google.

Other business reactions to the GDPR demonstrate some interesting and unexpected consequences for users. Chinese tech giant WeChat (owned by Tencent) purports to apply GDPR standards for international users of the service—but not for Chinese users.[100] This can be contrasted with Facebook, which no longer applies EU data protection law standards to its non-EU users outside the EU, the United States, and Canada. Prior to the GDPR's coming into force, these Facebook users fell within the competence of Facebook's Ireland office and were subject to Irish law, which incorporates EU laws such as the GDPR; now they are subject to the US terms and conditions, involving weaker privacy standards.[101] Thus the position for Facebook's non–North American, non-EU users is now worse in terms of their privacy protection than it was prior to the GDPR's implementation and is also weaker than the data protection standards WeChat would apply to them. Withdrawing services from the EU is another business reaction to the GDPR: some US-based news publications have blocked EU-based users from accessing their websites rather than complying with the GDPR.[102] Instead, users are redirected to a page stating that the publication is not available in the EU.

Thus the GDPR is an example of the Brussels Effect, operating as a legal innovation exporting the EU's soft power in the multipolar internet and maintaining the EU's relevance as a global economic power, rather than improving the competitiveness of the EU's own internet sector. However, there are unintended consequences of this for transnational Big Tech corporations, such as the reinforcement of Google's power within the EU, and an attempt by firms such as Facebook to stem the GDPR's application to users beyond the EU, whereas Chinese-based WeChat claims to apply these standards to non-EU, non-Chinese users. Aspects of the GDPR's substance exhibit desirable goals for privacy protection and other user interests, but ultimately the economic aspects of the GDPR are still grounded in a capitalist system and still permit data collection in many cases.

Conclusion

Does the GDPR demonstrate a genuine concern on the EU's part with surveillance and neoliberal digital capitalism by protecting and promoting users' privacy and data protection rights? Certainly, there are aspects of the GDPR that indeed do this. However, there are other aspects of the GDPR, which in their form and in their effect, facilitate capitalist markets and activities involving personal data, and defer some governance activities to the private sector. Thus, the GDPR accords with Halpin and Simpson's description of EU internet regulation as "mixed mode" with neoliberal aspects and interventionalist aspects aimed at social goals, here preserving human rights, especially privacy.[103] The GDPR is, in part, a form of "regulatory surveillance capitalism" that sets out a framework for, and some limits on, the collection of personal data but does not stop this, nor does it fundamentally challenge the datafied Big Tech status quo.

The GDPR as part of the EU's data protection law and policy more generally may also pursue certain commercial goals in the context of an increasingly multipolar internet governance scenario, by "exporting" the EU model of data protection internationally via the Brussels Effect through the GDPR's territorial reach and in EU trade policy. However, it seems this Brussels Effect is largely one that promotes the EU's norms and soft power more than its own internet industry if Google's continued and reinforced power in the EU is indicative of wider trends. The GDPR's complex and conflicting aims and effects do not permit a simple determination or verdict on the law's relationship with surveillance capitalism and the EU's own digital industry. However, this complexity may point to the GDPR as an instrument of regulatory (surveillance) capitalism whereby there is some curbing of the excesses of Big Tech and datafication but the underlying political economy is largely left undisturbed.

Ultimately, the GDPR's "proof" will be in its continued implementation and enforcement. However, the reinforcement of Google's market position through

GDPR compliance demonstrates aspects of Birnhack and Elkin-Koren's "invisible handshake" whereby regulation, including in the internet sphere, may benefit large players by excluding smaller players from the market due to the costs of compliance.[104] Some (unintended) consequences of the GDPR may be a regulatory surveillance capitalist internet "with European characteristics," these characteristics being better data protection compliance and oversight than in other parts of the world—yet, Big Tech, whether from the United States or China, still collects EU residents' data, albeit in GDPR compliant ways.

In this sense, the GDPR fails to ignite postcapitalist and post-globalization trends alike. There may be a further reinforcement of (US) Big Tech in the COVID-19 digital response, whereby Google and Apple are positioning their smartphone tracking software as a data protection compliant "Privacy Preserving Contact Tracing" option, "outflank[ing]" European governments' own initial attempts to develop contact tracing apps.[105] Data protection compliance may be a tool to further reinforce large (US) tech companies' market power.

Ultimately, the GDPR does not make a fundamental break with previous EU internet regulation, which demonstrated a mixture of neoliberal aspects and the pursuit of specific social and commercial goals. In doing so, it left "gaps" whereby individuals' rights and interests were not adequately protected vis-à-vis the state and large internet companies—gaps that seem to persist after the GDPR.[106] Data protection law merely entails that practices of datafication can still occur by both public and private actors in ways that are compliant with the regulation. Instead of a fundamental challenge to the surveillance capitalist practices of Big Tech companies, the GDPR provides some limitations on such practices but does not tackle the overall power of Big Tech and its influence over individuals' and communities' data. Furthermore, the GDPR does not fully contribute to EU internet sovereignty in practice given that significant power still resides with non-EU companies such as Google—power that the GDPR seems to reinforce rather than weaken.

However, the GDPR cannot be expected to address the full gamut of issues posed by the transnational digital economy and society. Antitrust/competition law and consumer law in jurisdictions including the EU may constrain, in part, the power of Big Tech. Even with the Brussels Effect, law and regulation is a necessary but not sufficient tool in creating and sustaining a more just digital economy and society. Technical affordances are also paramount: the GDPR does acknowledge, for instance, the need for the technical embedding of norms contained in Article 25 on data protection by design and by default. One major challenge for the success of any legal or regulatory attempt to constrain Big Tech and datafication is the current scenario in which datafied practices are prevalent and Big Tech companies are very large and powerful.

Imagining and implementing postcapitalist alternatives, which regulation and policy can facilitate by creating the right conditions for innovative development in sustainable and equitable ways, must be a key goal for different stakeholder groups and activists in the EU and elsewhere.[107] Those within the EU must also resist the Brussels Effect of the GDPR becoming a neocolonial measure as critiqued by Arora,[108] and be cognizant of the limits of the EU (digital) project in how it furthers historical colonialism by "othering" those on its borders, especially those trying to enter, as we can see in Halkort's contribution to this volume.[109] Instead, we need a truly cosmopolitan and inclusive approach to technology and innovation governance for data protection. Whether EU law and policy can assist in creating these conditions in the EU, whether this is enough to counter the power of surveillance capitalism from either the United States or China, or whether the EU will align with the US against China on realpolitik "national security" grounds,[110] is something that remains to be seen in the coming years.

NOTES

1. Heikki Patomäki, "Neoliberalism and the Global Financial Crisis," *New Political Science* 31, no. 4 (2009): 431–42.
2. See, e.g., Sadık Kılıç, "Does COVID-19 as a Long Wave Turning Point Mean the End of Neoliberalism?" *Critical Sociology* 47, no. 4–5 (2021); Alfredo Saad-Filho, "From COVID-19 to the End of Neoliberalism," *Critical Sociology* 46, no. 4–5 (2020): 477–85.
3. Paramjit Singh, "Beyond the COVID-19 Pandemic: Gauging Neoliberal Capitalism and the Unipolar World Order," *International Critical Thought* 10, no. 4 (2020): 635–54.
4. Brett Aho and Roberta Duffield, "Beyond Surveillance Capitalism: Privacy, Regulation and Big Data in Europe and China," *Economy and Society* 49, no. 2 (2020): 187–212.
5. See, e.g., Giovanni Buttarelli, "The EU GDPR as a Clarion Call for a New Global Digital Gold Standard," *International Data Privacy Law* 6, no. 2 (2016): 77–78.
6. Haim Sandberg, "What Is Legal Innovation?" *University of Illinois Law Review* Online (2021): 63–76, https://illinoislawreview.org/online/what-is-legal-innovation/.
7. Anu Bradford, *The Brussels Effect: How the European Union Rules the World* (Oxford: Oxford University Press, 2020); Anu Bradford, "The Brussels Effect," *Northwestern University Law Review* 107, no. 1 (2012): 1–68, 3.

8. Bradford, *The Brussels Effect*, 23.
9. David Levi-Faur, "The Global Diffusion of Regulatory Capitalism," *Annals of the American Academy of Political and Social Science* 598 (2005): 12–32.
10. Paul Mason, *PostCapitalism: A Guide to Our Future* (London: Allen Lane, 2015); Nick Srnicek and Alex Williams, *Inventing the Future: Postcapitalism and a World without Work* (London: Verso, 2015).
11. Bradford, *The Brussels Effect*, 24.
12. Matthew Feeley, "EU Internet Regulation Policy: The Rise of Self-Regulation," *Boston College International and Comparative Law Review* 22, no. 1 (1999): 159–74, 167.
13. Feeley, "EU Internet Regulation Policy."
14. Feeley, "EU Internet Regulation Policy."
15. George Christou and Seamus Simpson, "The Internet and Public–Private Governance in the European Union," *Journal of Public Policy* 26, no. 1 (2006): 43–61. See also Christopher Marsden, *Internet Co-Regulation: European Law, Regulatory Governance and Legitimacy in Cyberspace* (Cambridge: Cambridge University Press, 2011).
16. George Christou and Seamus Simpson, "The European Union, Multilateralism and the Global Governance of the Internet," *Journal of European Public Policy* 18, no. 2 (2011): 241–57.
17. Marsden, *Internet Co-Regulation*, 46.
18. Edward Halpin and Seamus Simpson, "Between Self-Regulation and Intervention in the Networked Economy: The European Union and Internet Policy," *Journal of Information Science* 28, no. 4 (2002): 285–96.
19. Angela Daly, *Private Power, Online Information Flow and EU Law: Mind the Gap* (Oxford: Hart, 2016).
20. Halpin and Simpson, "Between Self-Regulation and Intervention in the Networked Economy."
21. David M. Kotz, "The Financial and Economic Crisis of 2008: A Systemic Crisis of Neoliberal Capitalism," *Review of Radical Political Economics* 41, no. 3 (2009): 305–17.
22. Mason, *PostCapitalism*; Srnicek and Williams, *Inventing the Future*.
23. See Paul Chatterton and Andre Pusey, "Beyond Capitalist Enclosure, Commodification and Alienation: Postcapitalist Praxis as Commons, Social Production and Useful Doing," *Progress in Human Geography* 44, no. 1 (2020): 27–48.
24. Patomäki, "Neoliberalism and the Global Financial Crisis"; Engelbert Stockhammer, "Neoliberal Growth Models, Monetary Union and the Euro Crisis. A Post-Keynesian Perspective," *New Political Economy* 21, no. 4 (2016): 365–79.
25. Nick Srnicek, "The Challenges of Platform Capitalism: Understanding the Logic of a

New Business Model," *Juncture* 23, no. 4 (2017): 254–57.

26. Rainer Mühlhoff, "We Need to Think Data Protection beyond Privacy: Turbo-Digitalization after COVID-19 and the Biopolitical Shift of Digital Capitalism," *Medium*, March 30, 2020, http://dx.doi.org/10.2139/ssrn.3596506; Christian Fuchs, "Everyday Life and Everyday Communication in Coronavirus Capitalism," *tripleC* 18, no. 1 (2020): 375–99.

27. Halpin and Simpson, "Between Self-Regulation and Intervention in the Networked Economy."

28. Sophie Robin-Olivier, "The 'Digital Single Market' and Neoliberalism: Reflections on Net Neutrality," in *Legal Trajectories of Neoliberalism: Critical Inquiries on Law in Europe*, ed. Margot E. Salomon and Bruno De Witte (Florence: Robert Schuman Centre for Advanced Studies Research Paper No. RSCAS 2019/43, 2019), 45–49, http://dx.doi.org/10.2139/ssrn.3425753.

29. David Lyon, "The Snowden Stakes: Challenges for Understanding Surveillance Today," *Surveillance & Society* 13, no. 2 (2015).

30. Yong Jin Park and Marko Skoric, "Personalized Ad in Your Google Glass? Wearable Technology, Hands-Off Data Collection, and New Policy Imperative," *Journal of Business Ethics* 142 (2017): 71–82.

31. Madeline Carr, "Internet Freedom, Human Rights and Power," *Australian Journal of International Affairs* 67, no. 5 (2013): 621–37. Shawn M. Powers and Michael Jablonski, *The Real Cyber War: The Political Economy of Internet Freedom* (Champaign: University of Illinois Press, 2015); Blayne Haggart and Michael Jablonski, "Internet Freedom and Copyright Maximalism: Contradictory Hypocrisy or Complementary Policies?" *Information Society* 33, no. 3 (2017): 103–18.

32. Shoshana Zuboff, *The Age of Surveillance Capitalism: The Fight for a Human Future at the New Frontier of Power* (London: Profile Books, 2019).

33. Michael Birnhack and Niva Elkin-Koren, "The Invisible Handshake: The Reemergence of the State in the Digital Environment," *Virginia Journal of Law & Technology* 8, no. 2 (2003): 1–57.

34. Daly, *Private Power, Online Information Flow and EU Law*.

35. Lorna Woods, "ICO Reacts to Use of Data Analytics in Micro-Targeting for Political Purposes," *European Data Protection Law Review* 4, no. 3 (2018): 381–83.

36. Carole Cadwalladr, "The Great British Brexit Robbery: How Our Democracy Was Hijacked," *The Guardian*, May 7, 2017; Roberto González, "Hacking the Citizenry? Personality Profiling, 'Big Data' and the Election of Donald Trump," *Anthropology Today* 33 (2017): 9–12; Tom Dobber, Ronan Ó Fathaigh, and Frederik Zuiderveen Borgesius,

"The Regulation of Online Political Micro-Targeting in Europe," *Internet Policy Review* 8, no. 4 (2019).

37. Aho and Duffield, "Beyond Surveillance Capitalism."
38. Aho and Duffield, "Beyond Surveillance Capitalism."
39. Aho and Duffield, "Beyond Surveillance Capitalism."
40. Xiao Qiang, "The Road to Digital Unfreedom: President Xi's Surveillance State," *Journal of Democracy* 30, no. 1 (2019): 53–67.
41. See Elisabeth Eide, "Chilling Effects on Free Expression: Surveillance, Threats and Harassment," in *Making Transparency Possible: An Interdisciplinary Dialogue*, ed. Roy Krøvel and Mona Thowsen (Oslo: Cappelen Damm Akademisk, 2019), 227–42; Jonathon Penney, "Internet Surveillance, Regulation, and Chilling Effects Online: A Comparative Case Study," *Internet Policy Review* 6, no. 2 (2017), https://doi.org/10.14763/2017.2.692.
42. Shaojung Sharon Wang and Junhao Hong, "Discourse behind the Forbidden Realm: Internet Surveillance and Its Implications on China's Blogosphere," *Telematics and Informatics* 27, no. 1 (2010): 67–78, https://doi.org/10.1016/j.tele.2009.03.004.
43. Daly, *Private Power, Online Information Flow and EU Law*.
44. See Julian Gruin, "Financializing Authoritarian Capitalism: Chinese Fintech and the Institutional Foundations of Algorithmic Governance," *Finance and Society* 5, no. 2 (2019): 84–104.
45. Rosemary Segurado, "The Brazilian Civil Rights Framework for the Internet: A Pioneering Experience in Internet Governance," in *The Internet and Health in Brazil*, ed. Andre Pereira Neto and Matthew Flynn (Cham, Switzerland: Springer 2019), 27–46.
46. Francis Augusto Medeiros and Lee Bygrave, "Brazil's *Marco Civil da Internet*: Does It Live Up to the Hype?" *Computer Law & Security Review* 31, no. 1 (2015): 120–30.
47. Segurado, "The Brazilian Civil Rights Framework for the Internet."
48. Ron Deibert and Louis Pauly, "Mutual Entanglement and Complex Sovereignty in Cyberspace," in *Data Politics: Worlds, Subjects, Rights*, ed. Didier Bigo, Engin Isin, and Evelyn Ruppert (London: Routledge, 2019), 81–99.
49. See Julia Bader, "To Sign or Not to Sign: Hegemony, Global Internet Governance, and the International Telecommunication Regulations," *Foreign Policy Analysis* 15, no. 2 (2019): 244–62.
50. Dennis Broeders, Liisi Adamson, and Rogier Creemers, "Coalition of the Unwilling? Chinese and Russian Perspectives on Cyberspace," Hague Program for Cyber Norms, policy brief, November 2019.
51. Broeders, Adamson, and Creemers, "Coalition of the Unwilling?"; Eva Claessen,

"Reshaping the Internet—The Impact of the Securitisation of Internet Infrastructure on Approaches to Internet Governance: The Case of Russia and the EU," *Journal of Cyber Policy* 5, no. 1 (2020): 140–57; Yu Hong and G. Thomas Goodnight, "How to Think about Cyber Sovereignty: The Case of China," *Chinese Journal of Communication* 13, no. 1 (2020): 8–26.

52. Yu Hong, *Networking China: The Digital Transformation of the Chinese Economy* (Champaign: University of Illinois Press, 2017).
53. Aho and Duffield, "Beyond Surveillance Capitalism."
54. Mercy A. Kuo, "The Digital War: US–China Tech Competition in the Biden Era: Insights from Winston Ma," *The Diplomat*, January 26, 2021, https://thediplomat.com/2021/01/the-digital-war-us-china-tech-competition-in-the-biden-era/.
55. Michael Keane and Haiqing Yu, "A Digital Empire in the Making: China's Outbound Digital Platforms," *International Journal of Communication* 13 (2019): 4624–41; Yu Hong and Eric Harwit, "China's Globalizing Internet: History, Power, and Governance," *Chinese Journal of Communication* 13, no. 1 (2020): 1–7; Madison Cartwright, "Internationalising State Power through the Internet: Google, Huawei and Geopolitical Struggle," *Internet Policy Review* 9, no. 3 (2020); Lianrui Jia and Lotus Ruan, "Going Global: A Comparative Study of Chinese Mobile Applications' Data and User Privacy Governance at Home and Abroad," *Internet Policy Review* 9, no. 3 (2020).
56. Keane and Yu, "A Digital Empire in the Making."
57. Sophie Meunier and Kalypso Nicolaïdis, "The European Union as a Conflicted Trade Power," *Journal of European Public Policy* 13, no. 6 (2006): 906–25.
58. Peter Hall, "Varieties of Capitalism in Light of the Euro Crisis," *Journal of European Public Policy* 25, no. 1 (2018): 7–30.
59. Rajan Menon and William Ruger, "NATO Enlargement and US Grand Strategy: A Net Assessment," *International Politics* 57 (2020): 371–400.
60. Didier Bigo, "Beyond National Security, the Emergence of a Digital Reason of State(s) Led by Transnational Guilds of Sensitive Information: The Case of the Five Eyes Plus Network," in *Research Handbook on Human Rights and Digital Technology: Global Politics, Law and International Relations*, ed. Ben Wagner, Matthias C. Kettemann, and Kilian Vieth (Cheltenham, UK: Edward Elgar, 2019), 33–52.
61. Patrick F. Walsh and Seumas Miller, "Rethinking 'Five Eyes' Security Intelligence Collection Policies and Practice Post Snowden," *Intelligence and National Security* 31, no. 3 (2016): 345–68.
62. Zygmunt Bauman, Didier Bigo, Paulo Esteves, Elspeth Guild, Vivienne Jabri, David Lyon, and R. B. J. Walker, "After Snowden: Rethinking the Impact of Surveillance,"

International Political Sociology 8, no. 2 (2014): 121–44.

63. Jean Paul Simon, "How Europe Missed the Mobile Wave," *info* 18, no. 4 (2016): 12–32, https://doi.org/10.1108/info-02-2016-0006; Harald Gruber, "Proposals for a Digital Industrial Policy for Europe," *Telecommunications Policy* 43 no. 2 (2019): 116–27.

64. Michael Kwet, "Digital Colonialism: US Empire and the New Imperialism in the Global South," *Race & Class* 60, no. 4 (2019): 3–26; Smith Mehta, "Localization, Diversification and Heterogeneity: Understanding the Linguistic and Cultural Logics of Indian New Media," *International Journal of Cultural Studies* 23, no. 1 (2020): 102–20; Marcos Vinícius Isaias Mendes, "The Limitations of International Relations Regarding MNCs and the Digital Economy: Evidence from Brazil," *Review of Political Economy* 33, no. 1 (2021); Hong Shen, "Building a Digital Silk Road? Situating the Internet in China's Belt and Road Initiative," *International Journal of Communication* 12 (2018): 2683–701; see Lianrui Jia and David Nieborg in this volume.

65. Angela Daly, Thilo Hagendorff, Li Hui, Monique Mann, Vidushi Marda, Ben Wagner, and Wayne Wei Wang, "AI, Governance and Ethics: Global Perspectives," in *Constitutional Challenges in the Algorithmic Society*, ed. Oreste Pollicino and Giovanni de Gregorio (Cambridge: Cambridge University Press, 2021).

66. Claessen, "Reshaping the Internet."

67. European Commission, *Shaping Europe's Digital Future* (2020), https://ec.europa.eu/info/strategy/priorities-2019-2024/europe-fit-digital-age/shaping-europe-digital-future_en.

68. European Commission, *Shaping Europe's Digital Future*.

69. Terry Flew, "Globalization, Neo-globalization and Post-globalization: The Challenge of Populism and the Return of the National," *Global Media and Communication* 16, no. 1 (2020): 19–39.

70. Srikara Prasad, Malavika Raghavan, Beni Chugh, and Anubhutie Singh, "Implementing the Personal Data Protection Bill: Mapping Points of Action for Central Government and the Future Data Protection Authority in India," Dvara Research, policy brief (2019), https://ssrn.com/abstract=3521132.

71. Menaka Guruswamy, "Justice K.S. Puttaswamy (Ret'd) and Anr v. Union of India and Ors," *American Journal of International Law* 111, no. 4 (2017): 994–1000.

72. See Lu Yu and Björn Ahl, "China's Evolving Data Protection Law and the Financial Credit Information System: Court Practice and Suggestions for Legislative Reform," *Hong Kong Law Journal* 51, no. 1 (2021).

73. George Qi, Qianqian Li, Gretchen Ramos, and Darren Abernethy, "China Releases Draft Personal Information Privacy Law," *National Law Review* 11, no. 71 (March 12, 2021).

74. Bo Zhao and Yang Feng, "Mapping the Development of China's Data Protection Law: Major Actors, Core Values and Shifting Power Relations," *Computer Law & Security Review* 40 (April 2021).
75. Leonardo Parentoni and Henrique Cunha Souza Lima, "Protection of Personal Data in Brazil: Internal Antinomies and International Aspects," *International Conference on Industry 4.0 and Artificial Intelligence Technologies—INAIT 2019* (2019), http://dx.doi.org/10.2139/ssrn.3362897; Marcos Viana da Silva, Erick da Luz Scherf, and Jose da Silva, "The Right to Data Protection versus 'Security': Contradictions of the Rights-Discourse in the Brazilian General Personal Data Protection Act (LGPD)," *Revista Direitos Culturais* 15, no. 36 (2020): 209–32, https://doi.org/10.20912/rdc.v15i36.18; Jeff Kuo, "Brazil Postpones Enforcement of New Privacy Law in Response to COVID-19," *Lexology*, May 4, 2020, https://www.lexology.com/library/detail.aspx?g=6d3b2eff-948c-4879-be6a-267413e5d68b.
76. Tetyana Lokot, "Data Subjects vs. People's Data: Competing Discourses of Privacy and Power in Modern Russia," *Media and Communication* 8, no. 2 (2020): 314–22, 316.
77. Alfred Mavedzenge, "The Right to Privacy v. National Security in Africa: Towards a Legislative Framework Which Guarantees Proportionality in Communications Surveillance," *African Journal of Legal Studies* 12, no. 3–4 (2020): 360–90; Lukman Adebisi Abdulrauf, "Data Protection in the Internet: South Africa," in *Data Protection in the Internet*, ed. Dario Moura Vicente and Sofia de Vasconcelos Casimiro (Cham, Switzerland: Springer, 2020), 349–70; Nerushka Bowan, "After 7-year Wait, South Africa's Data Protection Act Enters into Force," *IAPP Privacy Tracker*, July 1, 2020, https://iapp.org/news/a/after-a-7-year-wait-south-africas-data-protection-act-enters-into-force/.
78. Maja Brkan, "The Essence of the Fundamental Rights to Privacy and Data Protection: Finding the Way through the Maze of the CJEU's Constitutional Reasoning," *German Law Journal* 20, no. 6 (2019): 864–83; Christopher Docksey and Hielke Hijmans, "The Court of Justice as a Key Player in Privacy and Data Protection," *European Data Protection Law Review* 5, no. 3 (2019): 300–316
79. Svetlana Yakovleva and Kristina Irion, "Toward Compatibility of the EU Trade Policy with the General Data Protection Regulation," *AJIL Unbound* 114 (2020): 10–14.
80. Christian Peukert, Stefan Bechtold, Michail Batikas, and Tobias Kretschmer, "European Privacy Law and Global Markets for Data" (2020), https://ssrn.com/abstract=3560392.
81. Catherine Barrett, "Are the EU GDPR and the California CCPA Becoming the De Facto Global Standards for Data Privacy and Protection?" *Scitech Lawyer* 15, no. 3 (2019): 24–29.

82. Samson Yoseph Esayas and Angela Daly, "Proposed Australian Consumer Right to Access and Use Data: A European Comparison," *European Competition and Regulatory Law Review* 2, no. 3 (2018): 187–202.
83. Payal Arora, "GDPR—A Global Standard? Privacy Futures, Digital Activism and Surveillance Cultures in the Global South," *Surveillance & Society* 17, no. 5 (2019): 717–25.
84. Payal Arora, "Decolonizing Privacy Studies," *Television & New Media* 20, no. 4 (2019): 366–78.
85. Yakovleva and Irion, "Toward Compatibility."
86. Yakovleva and Irion, "Toward Compatibility"; Svetlana Yakovleva and Kristina Irion, "Pitching Trade against Privacy: Reconciling EU Governance of Personal Data Flows with External Trade," *International Data Privacy Law* 10, no. 3 (2020).
87. Yakovleva and Irion, "Pitching Trade against Privacy."
88. Yakovleva and Irion, "Pitching Trade against Privacy."
89. Orla Lynskey, *The Foundations of EU Data Protection Law* (Oxford: Oxford University Press, 2015).
90. Daly, *Private Power, Online Information Flow and EU Law*.
91. Wanshu Cong, "Privacy and Data Protection: Solving or Reproducing the Democratic Crisis of the Neoliberal Capitalism?," paper presented at Computers, Privacy & Data Protection Conference: Data Protection and Democracy, 30 January–1 February 2019, https://papers.ssrn.com/sol3/papers.cfm?abstract_id=3744917.
92. Iris Van Ooijen and Helena U. Vrabec, "Does the GDPR Enhance Consumers' Control over Personal Data? An Analysis from a Behavioural Perspective," *Journal of Consumer Policy* 42 (2019): 91–107.
93. Daly, *Private Power, Online Information Flow and EU Law*.
94. Tal Zarsky, "Incompatible: The GDPR in the Age of Big Data," *Seton Hall Law Review* 47 (2016): 995–1020.
95. Daly, *Private Power, Online Information Flow and EU Law*.
96. Irene Kamara, "Co-regulation in EU Personal Data Protection: The Case of Technical Standards and the Privacy by Design Standardisation 'Mandate,'" *European Journal of Law and Technology* 8, no. 1 (2017); Maximilian Grafenstein, "Co-Regulation and the Competitive Advantage in the GDPR: Data Protection Certification Mechanisms, Codes of Conduct and the 'State of the Art' of Data Protection-by-Design," in *Research Handbook on Privacy and Data Protection Law: Values, Norms and Global Politics*, ed. Gloria González-Fuster, Rosamunde van Brakel, and Paul De Hert (Cheltenham, UK: Edward Elgar Publishing, 2019).

97. Peukert, Bechtold, Batikas, and Kretschmer, "European Privacy Law and Global Markets for Data."
98. Peukert, Bechtold, Batikas, and Kretschmer, "European Privacy Law and Global Markets for Data."
99. Damien Geradin, Dimitrios Katsifis, and Theano Karanikioti, "GDPR Myopia: How a Well-Intended Regulation Ended Up Favoring Google in Ad Tech," TILEC Discussion Paper No. 2020-012 (2020), http://dx.doi.org/10.2139/ssrn.3598130.
100. Ines Casserly, "WeChat Reminds Users of Its Privacy Policy [Update]," *TNW*, September 23, 2017, https://thenextweb.com/news/wechat-reminds-users-of-its-privacy-statement-update; Jia and Ruan, "Going Global."
101. Alex Hern, "Facebook Moves 1.5bn Users out of Reach of New European Privacy Law," *The Guardian*, April 19, 2018.
102. Alex Hern and Martin Belam, "LA Times among US-Based News Sites Blocking EU Users Due to GDPR," *The Guardian*, May 25, 2018.
103. Halpin and Simpson, "Between Self-Regulation and Intervention in the Networked Economy."
104. Birnhack and Elkin-Koren, "The Invisible Handshake."
105. Mark Scott, Elisa Braun, Janosch Delcker, and Vincent Manancourt, "How Google and Apple Outflanked Governments in the Race to Build Coronavirus Apps," *Politico*, May 15, 2020.
106. Daly, *Private Power, Online Information Flow and EU Law*.
107. See Angela Daly, S. Kate Devitt, and Monique Mann, eds., *Good Data* (Amsterdam: Institute of Network Cultures, 2019).
108. Arora, "Decolonizing Privacy Studies."
109. See Halkort in this volume.
110. Marc Rotenberg, "*Schrems II*, from Snowden to China: Toward a New Alignment on Transatlantic Data Protection," *European Law Journal* 26, no. 1–2 (2020): 141–52.

The Global versus the National

Creativity in Turkey's Game Industry

Serra Sezgin and Mutlu Binark

The digital game industry, which is rapidly growing around the world, can be considered one of the youngest and yet the most promising of all creative industries. Besides its economic value, its sociocultural impact and relevance lie in three parallel roots. First, many people around the world play digital games. Ever since mobile games became part of our lives, digital games have become far more accessible to, and consumed by, a wide range of people.

Second, compared to other creative industries, production barriers are much lower, for a small group of developers and even a single person can develop a digital game in a relatively short period of time. However, this does not mean that this market is completely free or has no barriers at all. According to Kerr, the largest markets are in North America, Western Europe, and parts of Asia.[1] In addition, a few highly competitive dominant companies, mostly located in North America, Japan, Korea, and China, are shaping the terms and conditions of market access. ElectronicArts, Blizzard, Sony, and Tencent actively publish and distribute these games, while simultaneously acting as the market's key gatekeepers. Moreover, their activities are often shaped not just by local cultures and tastes, but also by local and national policies and laws.[2] Hence, this new industry has boundaries regarding the market, region, content, and more.

The third reason for the sociocultural relevance of the digital game industry is that game developers, in their capacity as creative, cultural, or cognitive workers, possess what is deemed the most valuable cultural capital of our era: technical skills, and innovative and creative thinking capacity, as well as privileged educational backgrounds.[3] They are much sought after in the capitalist knowledge-based economy, because in addition to their willingness to take highly individualized risks and to be entrepreneurial, they can produce value by using their creativity.

In today's knowledge-based economy, innovation is an economic term with a human orientation (in the sense of human capital), while creativity forms a resource for profit maximizing and competitive advantage. As the introduction to this volume suggests, the critique of multipolar innovation and the discussion about the future of communication innovation require local and national perspectives as much as global angles. Developing countries such as Turkey, influenced by major actors like China and the United States, form a testing ground for the thesis of multipolar innovation and its effects on social transformation impacting labor, industries, and markets. Through such critical inquiry, the case of Turkey helps to unpack the politics, ethics, and struggles of multipolar communication innovation.

In this study, we focus mostly on the game developers working in the Middle East Technical University (METU) Technopark and Animation Technologies and Game Development Center (ATGC) by examining their perception of creativity and innovation. We then analyze speeches addressing these two issues by President Recep Tayyip Erdoğan, the leader of the ruling Justice and Development Party (JDP),[4] which has been the most powerful agent since the early 2000s in Turkish politics. Throughout this analysis, our discussion of these contradictory perceptions of creativity and innovation will uncover how the president's national-oriented policy affects this industry. The government's policy and regulations along with the game developers' perceptions outline a discursive field of struggle over the aforementioned sociocultural significance and value of the creative industries. Exploring the contradictions between the perception of creativity and innovation among game developers in Turkey, on the one hand, and President Erdoğan, on the other, may provide an opportunity for discussing the different articulations of concepts such as creativity and innovation in tandem with constructions of the game industry as either global or national in nature.

After summarizing this industry's brief history in Turkey and the establishment of METU Technopark (METUTP) and ATGC, we will introduce the findings of our field study on how the creative cluster's game developers perceive creativity and innovation. To understand this, we conducted in-depth interviews with twenty indigenous game developers from 2017 to 2018 and carried out nonparticipant observations (e.g., time spent in the field, observing the participants while working and attending professional and social events in METUTP). This latter methodology gave us a broader perspective on how they viewed their job and understood creative labor. Additionally, with the help of nonparticipant observation, we compared the interviews' findings with the participants' statements in informal/social environments. Thus, our final findings represent a combination of both interviews and nonparticipant observation.

The participants are workers at digital game studios, studio owners or directors as well as ex-workers, freelancers, or project partners. Employed by nine game studios, they are aged between 24 and 47 years old and have four to sixteen years of work experience. Four out of the twenty of them are women. All interviews were conducted in the field, mostly in the participants' working spaces or in public areas inside METUTP. Table 1 shows their ages and gender as well as the game company at which they work. The companies' names are represented by a letter to keep them anonymous; F represents freelance work.

From Amateur Phase to Institutionalized Phase

Creative industries with remarkable growth rates play a crucial role in the new economy and have been among the fastest growing sectors of the global economy worldwide.[5] For example, leisure, entertainment, media, and communication now represent 25 percent of the US economy. In fact, entertainment has displaced defense as Southern California's largest economic sector.[6] All governments of the Organisation for Economic Co-operation and Development (OECD) countries, including those of emerging economies such as Malaysia, are attempting to advance knowledge-based economy models based on competitive advantage. They do so by integrating strategies pertaining to the labor force, education, technology, and investment strategies, following the example of countries such as Japan, Singapore, and Finland. Thus, they are formulating industry policies that prioritize innovation and R&D-driven industries while reskilling and educating the population, as well as focusing on universalizing the benefits of connectivity by upgrading mass information and communication technology (ICT) literacy.[7]

Elisabeth C. Economy underlines that "China's leaders believe innovation is the key to their economic future. . . . The country's strategy to overcome its

TABLE 1. The Field Study Participants

PARTICIPANT	GENDER	AGE	COMPANY	JOB
Participant 1	M	29	X	Graphic Design
Participant 2	F	29	X	Software Development
Participant 3	M	28	X	Graphic Design
Participant 4	M	33	Y	Co-founder/Manager
Participant 5	F	24	F	Graphic Design
Participant 6	M	28	T	Co-founder/Manager
Participant 7	M	47	N	Co-founder/Manager
Participant 8	M	30	D	Software Development
Participant 9	M	34	D	Co-founder/Manager
Participant 10	F	32	G	Graphic Design
Participant 11	M	26	X	Software Development
Participant 12	M	35	M	Co-founder/Manager
Participant 13	M	29	Z	Graphic Design
Participant 14	F	30	X	Quality Testing
Participant 15	M	36	N	Graphic Design
Participant 16	M	26	N	Graphic Design
Participant 17	M	34	X	Graphic Design
Participant 18	M	38	N	Graphic Design
Participant 19	M	27	E	Software Development
Participant 20	M	29	Z	Co-founder/Manager

innovation gap is simple: spend on talent, spend on infrastructure, spend on research and development, and spend on others' technology."[8] China emphasizes the development of creative industries and creative labor forces as a driving force of techno-nationalist policy, while the "Chinese Dream" has been included in the CCP's ideological repertoire.[9] The Chinese Dream and its ideological repertoire emphasize the continuity of Chinese history, the magnificence of China's past, the commitment to the national development mission that will bring China to the "just" place it deserves in the world. Shaped by this discourse, China has its own new media ecosystem, as Lin and de Kloet as well as Jia and Nieborg explain in other chapters of this book, established by private enterprises under the auspices of the CCP and often referred to with the acronym BAT—Baidu, Alibaba, and Tencent.[10] Along with private investments, the Chinese state has funded its own "Silicon Valleys" as a techno-nationalist policy. Just like China, President Erdoğan has developed a techno-nationalist policy in Turkey, deploying a "National Will" discourse that is part of JDP's ideological repertoire, as will be explained later. Thus, Turkey has gradually invested in cultural and creative industries as a political tool via this policy.

Lazzeretti et al. indicate that Turkey's cultural and creative industries are highly concentrated in Istanbul and Ankara, where over 122,000 people (64 percent of total employment in this sector) are employed.[11] According to them, "software and programming was a fast-growing sector and reached more than 14% in both metropolitan centers."[12] Seçilmiş and Güran also underline that Ankara is the most dynamic hub for software programming and related creative industries.[13] In the early 2000s, small teams were developing digital games, mostly concentrated on localizing existing products; however, in 2008 Turkey started prioritizing its indigenous digital game industry as a part of the creative economy.[14]

Parallel to the growth of economic interest, investments in, and incentives for digital games, policymaking in this area has been accelerated since 2010. The increased incentives and public grants can be considered as primary actions to support the production of digital games. University–industry collaborations and governmental grants have encouraged young people to develop games.[15] In this section, we will briefly explain how this rapid increase of funding and incentives occurred, and we will introduce the main agents.

According to the Technology Development Zones Performance Index study conducted by the Ministry of Science, Industry and Technology between 2011 and 2018, METUTP has the highest performance of all technoparks and has ranked first for six years. This is hardly surprising, for most of the country's game studios, especially the small-scale ones and incubation centers, are located on its campus. Its location in Ankara also gives it physical proximity to governmental institutions and prestigious universities. Moreover, in 2008 the ATGC pre-incubation center was established there.

The Ministry of Development's 2013 report, *The Information Society Strategy Renewal Project: Information Technologies Sector Current Status Report*, states the importance of establishing techno-parks and incubation centers.[16] According to it, global companies dominate the software market, including Turkey's game industry.[17] However, ATGC is considered the most important pre-incubation center for young and creative entrepreneur candidates to realize their ideas.[18] For example, the report mentions projects carried out by seven groups founded within ATGC who subsequently became companies that enjoy the support of the Ministry of Industry and Trade.[19] The Ministry of Economy in particular decided to support the industry because of its potential to bring capital into the country. Therefore, game companies have received grants to carry out marketing activities, open offices abroad, and acquire consultancy services from abroad.[20]

The Ministry of Development's 2015 report, *The Information Society Strategy and Action Plan (2015–2018)*, is the most comprehensive policy paper on these creative industries so far. This action plan indicated that digital game culture is widespread in Turkey due to the country's young population, and proposed to convert the domestic market capacity to export.[21] In fact, a special article on the "creation of game industry" listed responsible institutions, among them METUTP; the Ministry of Economy; the Ministry of Science, Industry and Technology; the Ministry of Development; and the Ankara Development Agency.[22] The reasons for this investment are as follows: not only are game technologies intensively used in defense, health, and education sectors, but local games based on the "local" culture and history will contribute to "the presentation of our country."[23]

METUTP, which was authorized by this plan to determine future strategies for the industry, has been collaborating with ATGC to carry out this mission. Since its establishment in 2008, ATGC has accepted 150 game development teams, 22 of which have started their own companies; 95 percent of the rest have been integrated into different digital game companies.[24] Until today, more than 550 digital games have been developed and over one thousand people have attended training programs.

ATGC encourages amateur game developers to work in game studios and to establish their own companies in the future by providing hardware and an interactive working space, and organizing trainings focused on game production and business development. In 2016, its teams' export of digital games amounted to US$2 million.[25] Thus, ATGC plays a central role in the country's game industry within the context of both generating human capital and developing game projects. METUTP and ATGC continue to create various opportunities for start-ups to join the digital game ecosystem.[26]

As of this writing, the exact number of game companies, developers, or digital games produced in Turkey is unknown. However, the related associations have 61 member companies. When nonmember companies are added, the total estimated number is held to be around 80.[27] Despite this relatively low number, the development of an indigenous game industry has come a long way over the last decade; its exports exceeded US$1 billion in 2018 and its import rate has increased every year since 2013. (The annual export rates of Turkey's digital game industry in 2013: US$379m; 2014: US$395m; 2015: US$418m; 2016: US$500m; 2017: US$700m; 2018: US$1.05 billion).[28]

As of 2019, the top five countries in this industry, in terms of earnings, are China (US$23,198b), the United States (US$17,832b), Japan (US$11,626b), the United Kingdom (US$3,189b), and South Korea (US$3,034b).[29] Although those numbers don't show the exact shares in the global market per year, by looking at the top five countries' global revenue, it can be stated that Turkey is still a small actor considering the numbers. Although still a small player, Turkey's game industry revenue has grown considerably, as has the development of a game culture in terms of both production and consumption. That is why we locate its game industry in a more institutionalized, as opposed to an amateur, phase, and it is therefore important to examine the motivation of game developers and the government's view of the game industry as part and parcel of its Information Society strategy and also cultural policy.

The Perception of Creativity and Innovation While Working at Creative Clusters

Creative clusters like METUTP are heavily invested in creativity and innovation, but the particular articulations and meanings of these discourses demand further attention. Since the European Middle Ages, the concept of creativity has generally been used to refer to God and His creations. In the eighteenth century it defined a human characteristic or skill; in the nineteenth century it acquired a creative intelligence, authenticity, and transcendental value by becoming associated with the works of painters and writers. After the Second World War it began to be associated with people outside the art world, and today it has a strong bond with continuous change, innovation, and flexibility.[30]

Since creativity is an ambiguous and non-measurable concept, it is valuable as a disciplinary and regulatory discourse for capitalism.[31] In other words, who is "creative" and how this is determined are rather debatable questions, and thus are answered by the owners of capital according to their own needs. Kim suggests that, in addition to the ambiguous valuation of creativity in the labor process, more attention should be paid to the question of in what ways human, embodied knowledge and skills together with information and social capital have become main drivers and features of creativity.[32] Given that such producers contribute part of themselves to their products, the affective dimension renders the labor process more complex in the sense that separating the product from the laborer becomes difficult, and the motivations and expectations of work differ from those of traditional work.

Embedded in capital but always exceeding it, creativity is the common ability for social discovery and collaboration.[33] Its appropriation by neoliberal ideology positions information and personal skills as a commodity, thereby turning creativity into an economic competence that is tamed through educational and corporative management, instead of a common good directed toward social collective work and democratization.[34]

On the other hand, in terms of scientific and technological development, innovation as a concept became a technique to maximize profit. Gaining a competitive advantage over one's competitors requires that potential buyers be informed of its innovation(s). Innovation, as opposed to creation or invention, can be defined as "the process of economic change through the origination, adoption, and retention of new ideas into the economic order."[35] Given this, one can state that innovation has deeper roots within the economy and capitalist production than does creativity, for by its very nature it is an industrial, economic term. And yet like the concept of creativity, innovation has also gained a sense of a personal characteristic that determines both the product's and the individual's value.

In terms of the market, authentic/original products that are differentiated and adaptable to the economy—meaning innovative—provide a competitive advantage, just as an innovative individual who can think outside the box or has original ideas gains a competitive advantage in a neoliberal economy. The concept of creativity—a common ground for humanity—turns into an economic concept, and thus innovation, as an economic term, reflects on the valuing of individuals. Thus, the lines between a commodity and a person, as well as between economy and subjectivity, are blurred.

Our field research suggests that Turkey's game developers have embraced discourses of creativity and innovation that they relate to the global context of creative work, while they reject the more local and national adaptations of these terms. They consider themselves "global" creative workers and want to compete in the global labor market. According to our study's participants, the concept of creativity has two meanings that can be used either together or separately: (1) innovativeness and originality, and (2) inventive problem-solving skills revealed by thinking outside the box. The first definition points out the creative elements of a game as a product, whereas the latter one speaks to the developer's creativity as a skill or a talent.

The concept of creativity, seen in the context of innovation, involves being able to discover something that does not yet exist, as well as reinterpreting, reorganizing, or redeveloping an existing product via new methods. As one of our interviewees explained:

> [Participant 7]: Creativity is the key. Now there are millions of developers, games and downloaders. So, the outcome needs to be very good. In this sense, creativity means innovation. There are innovative works, but there is only a small chance to be the first as in the old days. Today, it is more common to adopt, mix, transform the successful games. So, it is possible to be creative without innovation.

As this participant explains, creativity includes but does not depend on innovation and, at some point, even exceeds it; other participants argue that a non-innovative product cannot be considered creative. This thought, which does not limit creativity to a product specialty, leads one to view creativity as a way of thinking and, therefore, a personal characteristic.

Specifically, our participants mentioned that there are three types of creativity in the development process: (1) the production of the creative idea during the design process, (2) the game's artistic quality in the audio and visual dimensions, and (3) the skills related to solving problems, which are needed at almost every level of the development process, including marketing and design, but primarily programming. Game developers indicate that they mostly use their creativity to figure out alternative methods to solve a problem. Participant 11 states that "this is not much different from solving a math problem"—finding a solution when others cannot. In this sense, creative problem-solving skills are part of a distinctive creativity for game developers.

Creativity in this industry also has an artistic dimension. Digital games are a promising and important part of the art and entertainment industry, just like film or music, that form and are formed by one's culture and society. In this sense, digital games' strong attachment to aesthetics and design has engendered an ongoing debate as to whether they can be considered an art form. Within the context of this study and drawing on an interactive art approach, this view foregrounds artistic elements and, in some cases, the intent of the developers' artistic expression.[36] Thus, capacity in the production process for their artistic and creative expression is essential to considering their labor and the creativity involved.

In other words, the development process involves creative production and thinking that, in most cases, allow developers to self-identify as creative people and, in some cases, as artists who have an opportunity to express themselves via this process. This also means that game production creates a possibility for developers to have a voice or to be heard by others, an important point that makes this job just as desirable as many other types of creative works that people produce out of romantic motivations like love and passion.[37] For instance, Participant 11 says, "If you add style (üslup) to the decision making process while developing a game, you turn your production into an artwork . . . I define art as a way of self-expression. I mean, if you make concessions to what is best in order to express yourself, it is art."

The artistic dimension of the work, as Tokumitsu indicates, makes workers feel like artists who create freely and with passion, unbound by any obligations.[38] As Participant 15 emphasized, the independent (indie) games produced with less financial and yet more artistic motivations especially express their creators' passion for creating something stunning and beautiful.

Our findings reveal that this artistic dimension plays a significant role in the developers' attitudes and perceptions of their jobs as well as the amount of labor expended. The more artistic and creative their game becomes, the more they feel like they are pursuing their dreams or expressing themselves instead of just working. Otherwise, the work becomes routinized and excludes creativity. Similarly, Bulut's research conducted in a game studio in America shows that the difference between fun and work disappears in the industry.[39]

However, the participants' opinions about games as an art form differ. Some interpret them as a holistic form of art that requires creativity on many levels, including design, planning, and marketing, and they define themselves as artists. Others consider games only partially as artistic products depending on the degree of artwork involved; or they argue that games are art only if they are developed with the intention of artistic expression and without any concern for financial gain. When a game is produced with commercial concern, or when it is developed by a bigger team (meaning that each developer only generates a small part of it), participants perceive the production process as less creative. As the artistic and creative nature of the production process decreases, the job seems more like a "real" job than a passion and form of self-expression.

Developers also regard creativity as a subjective quality, one that differs for each developer, while situating game development itself as an ever-changing and exciting creative job. Participant 14 underlines the need for creativity as follows:

> A game developer needs to be creative when developing the game because it is a very creative process. Sometimes you must find an idea that has never been thought before. The problem-solving ability needs to be very good, because it is impossible to find solutions if you can't look at it from different angles, especially when designing the game. Another factor where creativity comes into play is the artistic part of the work, [for] everything is created visually from the beginning. So, in general it is definitely a creative process already. So, the game companies cannot be too oppressive to employees. It's not something that can be produced under stress and pressure because it is not something mundane.

Participants stress the work's non-monotonous structure based on the assumption that the development process requires an innovative and creative approach. Although the digital game industry and the game ecosystem present themselves as spaces for innovative and creative thinking, our findings show that this is not always the case. As our participants indicate, the most important creativity-blocking factor is the routinization of work; others are stressful working conditions, political or economic settings, and self-censorship.

Consequently, our study observed that developers working in large-scale companies experience more frequently a feeling of alienation. In small-scale studios where approximately five or six developers work together, they are more involved in the creative thinking process and more connected to the game being developed. Additionally, the latter studios organize regular meetings at which they can share ideas and thus strengthen the sense of collectivity.[40] With the growth of capitalism, creativity became a personal asset, if not an individualist enterprise, rather than a communal dynamic of sharing and co-producing.[41] However, in our example, the participants' emphasis on community ties, communication, and the importance of a culture of solidarity in creative work point to the role of community dynamics in creativity and innovation.

Our interviews reveal that Turkey's game developers regard creativity and innovation in the context of the artistic, authentic, and innovative dimensions of product development. Remarkably, the participants mentioned neither these concepts' national or regional dimensions in terms of the economy, policy, or culture, nor their national or regional competitive advantages. Instead, their perception is more linked to individual and global qualifications, such as creativity and innovation, as these make the product and human capital globally competitive.

The JDP's Cultural Policy on Digital Games: The Emphasis on "National Essence"

In Turkey, information technologies and creative industries do not only fall within the scope of development policy and investments, but they are also recognized as cultural policy areas. Cultural policy has been an important area for Turkey's ideological battles since the very first days of the Turkish Republic, especially in terms of preserving, improving, and promoting the "national culture." Ada states that there was a strong—but unwritten—cultural policy during the Republican era, which lasted from 1920 to 2000, even though it was never stated in any document.[42] Deviating from the earlier republican, secularist orientation, JDP currently advances a new cultural policy revolving around symbolism connected with "Neo-Ottomanism," as we will explain shortly.[43] The JDP, which assumed power in 2002 as a self-defined "Conservative-Democrat" party,[44] has gradually moved away from liberal discourse and toward both an authoritarian and pro-Islamist governance style. Aksoy also points out that "Defining itself as a "conservative democrat" party, the JDP has now added a new layer to the official cultural policy that has so far been exclusively concerned with protecting national integrity."[45]

Today, the JDP constructs singular and homogenous cultural policy on being "national and local" ("milli ve yerli") based on the "national will" discourse.[46] Following the July 15, 2016, attempted coup d'état,[47] the government decided to isolate the country, especially in the cultural domain, by placing special emphasis on "national" culture.

According to President Erdoğan, the JDP has not been as successful in the realm of cultural policy as it would like to be. This is particularly true in the areas of media and popular culture, for the production and dissemination of Neo-Ottomanism have had the weakest impact there. Neo-Ottomanist discourse reinvents and reimagines the Ottoman legacy in modern Turkey.[48] Under the JDP regime, Neo-Ottomanism became a cultural policy, which constructs a new "national" identity through mass media, music, art, literature, and digital games as well.[49] Therefore, the various consecutive JDP governments have supported investments in the movie and television industry to create new images that construct the "national" in terms of the Neo-Ottomanist imagination, such as the movie *Fetih 1453* and the series *Diriliş Ertuğrul* (2014).[50] Furthermore, the Third National Culture Congress was held in 2017, and President Erdoğan has increasingly emphasized the "national essence" of cultural products, including innovation.[51]

Appadurai, who criticizes Erdoğan for turning culture into a theatre of sovereignty, contends that his "Neo-Ottomanism" cultural policy seeks to return to the Ottomans' traditions, language, and imperial glory both as a fantasy and a fetish.[52] According to him, Erdoğan and other populist authoritative leaders use cultural policy to justify their power, instead of opposing the exploitative system of neoliberalism and capitalism.[53] As one can see in his discursive practices, Erdoğan frequently establishes an "us-them" contrast through metaphors. By referring to stigmas, he polarizes the masses so that the national will, which he embodies in his own corporeality (existence), becomes a seemingly incontestable fact. The vocabulary that generally dominates his rhetoric encompasses phrases such as "one single nation, one single country, one single state," and "national will."[54]

In addition to producing national cultural programs, Erdoğan's speeches emphasize becoming an important economic power in the world. His perspective oscillates between situating Turkey in a global world and as a global power, on the one hand, and hailing localism, and nationalist and essentialist culture, on the other—constituting clearly a dichotomous and contradictory discourse. Here, we scan his various opening addresses and public speeches with a focus on the keywords of "creativity," "innovation," "creative industry," "information technologies," "industry 4.0," "internet," "social media," and "digital game" from April 28, 2014, to May 1, 2019 (see Table 2).[55] Obviously, how he frames these issues affects the perception of the digital game industry, creativity, and innovation in the field.

As detailed above, Erdoğan made twelve direct speeches on creative industries, the internet, and social media. When looking at them, we first notice an ongoing emphasis on the importance of developing national products and technologies, especially information technologies. In his speeches on the scope of "2023 Target," a policy agenda for the centennial anniversary of the Turkish Republic, the priority is to produce high added-value technological products. All the same, his rhetoric frequently refers to the importance of family and national values, as well as the dream of a big and strong Turkey for the future. He claims that globalization and the internet harm "national culture" and "values":

TABLE 2. President Erdoğan's Public Speeches, 2014–2019

DATE	TITLE OF THE SPEECH	THE SPEECH'S MAIN THEME	SUBTHEMES
11.11.2014	"One of the 2023 targets of Turkey is to be a productive country in information technology, not a consuming one"	Information technologies (ITs)	Producing national ITs
15.12.2014	The speech he delivered in TÜRKSAT 6A Local Communication Satellite Project Signature Ceremony	Information technologies	Producing national ITs
21.04.2015	The speech he delivered in the 175th Anniversary of Türk Telekom	Internet	The negative side effects of the use of internet
22.05.2015	"We are establishing a very strong and high-quality infrastructure in Education Through Fatih Project"	Information technologies	Producing national ITs
01.02.2017	"Every civilization produces its own technology, and every technology produces its own culture and value"	Information technologies	Producing national ITs
15.04.2018	"We have reached these days not through media operations, but by fighting against headlines"	Social media	The negative side effects of the use of social media
20.04.2018	"The biggest treasure of a nation is to have emotionally, intellectually and physically healthy generations"	Social media	The negative side effects of the use of internet
08.05.2018	"Local and national understanding should be our ideal in culture and arts like in everything"	Information technologies	Producing national ITs

DATE	TITLE OF THE SPEECH	THE SPEECH'S MAIN THEME	SUBTHEMES
11.05.2018	"We have challenged tutelage to leave a better future to our youth"	Social media	The negative side effects of the use of social media
24.01.2019	"We take our place in this process, saying the move is a national technology and digital Turkey"	Information technologies	Producing national ITs
06.02.2019	The speech delivered at the Opening Ceremony of METU Technopolis's Informatics and Innovation Center	Creativity and innovation	Producing national ITs
03.05.2019	Turkey's Innovation Week and Inovalig Awards Ceremony	Innovation	Producing national ITs

In an era when the world has been subject to gradual cultural desertification, only those societies that have a deeply rooted civilization and a civilization design aimed at the future can keep their originality. Those who fail in this regard will get lost among billions of societies. We are a nation with a very old civilization and we still preserve our strong civilization design. We are of course aware of the heavy damage done to us by media, communication, Internet, and popular culture.... The solution to defend and develop our civilization is not to fight against technology but use its opportunities.[56]

Erdoğan proposes the following antidote: "We have to rapidly overcome our failure in turning our culture-arts policies into the locomotive of our civilization design. A cultural and artistic atmosphere that is hostile to its own country, society, history and civilization, let alone supporting and pioneering it, will lead us nowhere but to submission to global popular culture. 'Local' and 'national' should be our ideals in culture and arts, just like in all other domains." His speech at the opening of the Third National Culture Congress,[57] held from March 3 to March 5, 2017, is especially important in terms of understanding his cultural policy and seeing how he constructs "we-others" in language:

> We need to rediscover and reconstruct our local and national cultural values against cultural alienation and cultural imperialism. That a cultural product has a local and national form never prevents it from having a universal meaning and message.... One of the biggest problems of our era is cultural flattening. No culture or civilization can be constructed with shallow works that are produced and consumed daily. We need to focus on permanent and long-term works. We should encourage especially our youth to learn real art and culture from real masters. My experience in political life also shows this. We must support and highlight the works that embrace our cultural wealth and maintain our values. We should not let television and especially social media devour our culture. On the contrary, we should seek ways to use these opportunities to pass our culture on to new generations.[58]

At this point, it is significant that Erdoğan regards cultural alienation and cultural imperialism as threats to "local and national culture." In addition to those nonlocal cultural productions and practices, those cultural products made in Turkey can also be "shallow."

Against this, Erdoğan proposes to spread "real" art and culture. Both broadcast media content considered unsuitable for JDP's conservative Islamist ideology, and the new media are labeled "evil," and they are portrayed as signs of "cultural flattening." The congress's concluding report announced the following motto: "Turkey for the Goodness of the World." In sum, the concluding report emphasizes that there is a need to protect Turkish culture from destructive threats, while advancing protective and essentialist cultural policies. Where these threats originate from, however, remains rather vague and ill-defined. By analyzing Erdoğan's speech at the opening of the congress, one can see that these perceived threats against the country's national and local culture potentially stem from anything that does not belong to "us." Culture is a field of political battle, and in Erdoğan's discourse, "real art and culture" is equal to "national and local essences." Given this understanding, he contends that both innovation and technology should have national and culturally "appropriate" values.

In his speech at the Fourth International Technology Addiction Congress, organized by the Green Crescent on November 27, 2017, Erdoğan said that technology should not be used against "creation,"[59] and added: "To us, the main criterion is to use technology to construct and develop the world, not to destroy it." He also associated information technology with internet addiction and indicated that children and young people should use such technology conscientiously. By emphasizing "creation," he appealed to a religious worldview and set of values.

During the opening ceremony of METUTP's Informatics and Innovation Center on February 6, 2019, the president's conceptualization of innovation was based primarily on technological developments that will improve Turkey's competitive strength and reproduce its national values. In his speech, he underlined technological independence and explained the basic features of "digital Turkey":[60]

> We must stand on our own feet in all areas from data production to data security, from defense, health and information technologies to artificial intelligence. If we cannot, others will be in control. We cannot maintain our independence without having a solid grasp of technology, just as we cannot be independent without ruling over our lands.... We cannot achieve our goals without becoming a country which designs, develops and produces technology as opposed to being a mere consumer. If the technology advances too fast, we need to move faster and work harder.

As seen above, President Erdoğan emphasizes the production of national technologies, and ties his argument to Western imperialism, particularly cultural imperialism. Based on his discursive strategy, then, he started the "Digital Transformation" project. In his speech on January 24, 2019, he added that he considers the issues highly important national matters, such as developing domestic and national innovative technologies, supporting the development of national software, and protecting critical infrastructures. He announced the establishment of the Digital Transformation Office with direct connection to his own office, and declared that "We also have a National Technology Movement and @DijitalTürkiye.[61] Thus the opportunities offered by science and technology will bring Turkey a big change."[62]

Recently, Erdoğan joined the Innovation Week Inovalig Awards Ceremony, held at the Istanbul Congress Center Turkey on May 3, 2019. In his speech, he said:

> We assume that digital transformation is a critical policy. We have accelerated the digital economy. The more we open to the world, the closer we get to our goals. Although some [countries] try to contain us, we will not fall into this trap.

He also mentioned innovation and its competitive advantages, saying that the exportation rates have increased. Although the speech was supposed to be about innovation and Innovation Week, it focused on conspiracy theories, among them such actors and subjects as the European Union, terror, other countries' future plans, or attacks on Turkey, as well as how "they" disapprove of "our" policy concerning Syrian immigrants. Also, he continued to reproduce the "us-them" discriminatory discourse throughout the potential threats, as understood from: "We are able to demonstrate a stronger stance against operations targeting our country. We were able to cut our own cord in every area against the attacks we were exposed to."

Contrary to the president's techno-nationalist and culturalist governance vision, our study's participants neither mention nor emphasize that their creative output should be "national" in nature; instead, they align creativity with more universalist values and a global context of collaboration and competition. However, following Erdoğan's public speeches, the Ministry of Youth and Sports has financed a strategy game, called *Nusrat* in 2015 (updated in 2018), and presented it as "national" and "local."[63] The game narrates the Battle of the Dardanelles during the First World War from the ideological perspective of the JDP. The ministry also established a database of so-called Islamophobic images in digital games with the purpose of denunciating these games. This mission of the ministry is well aligned with President Erdogan's techno-nationalist discourse and investment in "local" cultural production. Actions undertaken by this ministry form one way in which the president's discourse impacts the development of new software products and, furthermore, their public distribution.

To sum up, regarding digital games policy, the various JDP governments have been interested in informationalization and the IT industry since 2009. As a part of this industry, digital games have been considered a social, cultural, and political problem. Therefore, the JDP and Erdoğan have targeted them as an ideological battlefield, a realm that should be nationalized so that it accords with the "national will" discourse and Turkish-Islamic cultural values. The MFSP, the Ministry of Youth and Sports, and other ministries have been involved in the relevant policymaking and have urged public agents to develop age- and content-based regulations for this industry.

The government's framing of games and the game industry starkly contrasts with the framing by those working in this industry. A look at Turkey's game developers' perception reveals their belief that one can be "glocal," as opposed to "national," in creative industries. First, they have taught themselves mostly from tutorials prepared by globally successful or experienced developers. As many participants mention, "You need to improve yourself on your own, constantly, to be successful." Participants feel responsible for improving their own work, skills, and knowledge by looking beyond the national borders of Turkey since, as Participant 13 explains, "there is no one to teach, especially in Ankara; all successful people are in other countries." This attitude, which is widespread among the country's game developers, implies that the labor of game development is primarily individual and implicitly global. They see themselves as individual actors in the global game market and not as advocates of national cultural values. Indeed, our participants' future goals often included working in global companies located in other countries or developing independent games. To them, the "game market" implies the global market (i.e., global app stores and distribution channels such as Steam) and the global circuits of competition and collaboration in game development.

Conclusion: National Essence vs. Global Values

During the JDP's rule, and especially during the latest, prolonged state of emergency (2016–2018), freedom of expression and thought have been restricted in Turkey. Lawsuits are continuously being opened against artists and journalists for "insulting" the president. Thousands of academicians have been dismissed due to the decree-laws issued in August 2016.[64] However, creative workers in Turkey's game industry underlined the freedom of expression and thought during our field research. For them, these basic human values and rights are essential to being open-minded, creative, and original because creativity and entrepreneurship can only thrive where there is freedom of expression and thought, and innovative ideas can only sprout and grow in a soil of fertile ideas from multiple sources. Under such conditions, an individual with plenty of passion and imagination can think outside the box.

However, Erdoğan contends that information technologies should be used in line with the religious understanding of "creation" and emphasizes the need for the production of national technology in this direction. At the same time, contradictions appear in the commission reports published after the Third National Culture Congress. Although Erdoğan drew attention to the importance of the "national" and "local" in the country's cultural production, as well as to its current insufficiency, the Cultural Economy commission reports expressed the globality of the creative industry's productions. According to these reports, all the processes, starting with culture and creative industry content production to its sales and distribution, are both economic and political. Culture, a symbolic resource, cannot be considered separate from sociopolitical relations that implicate the "national" and "local" in transnational and global processes.[65]

In the above-cited speeches by Erdoğan, cultural production and creative industries are perceived as instruments for the JDP's Neo-Ottomanist cultural agenda, according to which "national" and "local" culture often means Sunni Islamic culture. Polo also emphasizes that the JDP's hegemonic power over and through the state apparatus could enable it to pressure artists and cultural actors.[66] Analyzing the various JDP governments' cultural policies reveals that they perceive the digital game industry as both local and national, whereas digital game developers often target the global market while developing a game and act as global actors in the game market. Moreover, they perceive creativity and innovation as existing on the individual or global level (usually in the context of artistic value), instead of within national or local frames.

But game developers in Turkey usually think of creativity and innovation together and approach them as concepts intrinsic to a specific human skill like thinking outside the box. This individualistic approach makes the workers responsible for developing their skills (or themselves) to be even more creative or innovative. But due to this ethos, these workers often fail to address the policymakers' responsibility to develop supportive policies and cultivate enabling work conditions and environments, such as by upholding the freedom of speech. Turkey's game developers rarely criticize the government, even if the oppressive political climate makes creative and innovative thinking difficult. The renowned Chinese artist-in-exile Ai Weiwei says that when Beijing talks about making Chinese culture strong and creative, it implements censorship and exterminates both individual thinking and the willingness to take risks and bear the ensuing consequences.[67] He stated: "It would be impossible to design an iPhone in China because it's not a product; it's an understanding of human nature."[68] In our opinion, a liberal cultural and democratic environment would do far more to improve Turkey's game industry than emphasizing nationality and locality in policy.

The opinions of Turkish game developers on censorship refer primarily to interpersonal communication with studio managers; however, they do not directly refer to Ankara's oppressive actions or policies. From this viewpoint, one can see that President Erdoğan's governance is distant from their perceptions and actions in their labor processes, whereas at some level its policies have actually had a negative impact on both of them.

Game developers also think that no state grants will be forthcoming if their productions do not align with the state's priorities.[69] This perceived concern has two negative implications: (1) the developers begin to employ self-censorship, and (2) the conflicting and different viewpoints within the government and public institutions will slow the industry's production and prevent its game companies and developers from having a role as cultural agents in the global market. As a result, creativity is constrained by the state's illiberal cultural policies and regulations. Moreover, given that the new bureaucratic and governance regime introduced in June 2018 has tied all the ministries to President Erdoğan's rule, we must analyze how his visions impact innovation, creativity, and his supposed goal of Turkey's digital transformation. At the 2019 Science Awards Ceremony of the Scientific and Technological Research Council of Turkey, he declared that "We are determined to become a country that produces new technologies and spreads them all over the World."[70]

> Countries that do not produce and use information and technology in the best way, can hardly work in the world of the future. With the knowledge we produce, we will contribute to the welfare of all humanity as well as reach our own goals. The aim of our National Technology Movement is to support this process.

The gaming industry in Turkey is particularly fascinating as a laboratory for competing views of innovation, creativity, and the relations between designers, markets, and the state. On the one hand, the decentralized nature of the gaming industry in Turkey, and the individualist perspectives of its designers, points to some of the key aspects of neoliberal capitalism as practiced in Silicon Valley. In both Silicon Valley's and Ankara's new tech corridors, we find game designers thinking of themselves as "artists," all while making products for corporations that reap millions of dollars of profit off this creativity. On the other hand, President Erdogan's Neo-Ottomanism and techno-nationalist policy, which stress the protection of some imagined indigenous Turkish cultural heritage against the onslaught of globalization, sound strikingly familiar to Chinese President Xi Jinping's rhetoric, and his national policy of "Chinese Dream" in which he positions China as an organic civilization buffeted by outside forces. Like China, Turkey wants to enjoy the benefits of globalized markets and communication innovations while controlling their impacts on the nation-state. Turkey's gaming industry is therefore positioned betwixt and between liberal market forces and national-oriented policy, seeking both global impacts via communication innovations and national protectionism. Obviously, the creative workers of METUTP's perception of developing technological products differ from President Erdoğan's. Therefore, global values and nationalist imaginaries will continue to engender conflicting discourses pertaining to technology production, particularly in Turkey's game industry.

NOTES

1. Aphra Kerr, *Global Games: Production, Circulation and Policy in the Networked Era* (New York: Routledge, 2017), 104–77.
2. Kerr, *Global Games*, 63.
3. Serra Sezgin, *Dijital Oyun Ekosistemi: Yaratıcı Endüstri ve Emek* (Ankara, Turkey: Alternatif Bilişim, 2020), 111–18, https://ekitap.alternatifbilisim.org/dijital-oyun-ekosistemi-yaratici-endustri-ve-emek/.
4. Recep Tayyip Erdoğan became the first elected president of the Turkish Republic in August 2014 and then he brought the "Presidential regime" to the public agenda. JDP (Justice and Development Party) was established on August 14, 2002, and its constitutive ideological components are nationalism, Turkish-Sunni Islamism, and patriarchal family structure. Its fundamental platform draws on populist right-wing discourse in compliance with neoliberal economic policies (Binark and Bayraktutan, "Discursive Strategies and Political Hegemony," 2018, 11–12).
5. Stuart Cunningham, "The Creative Industries after Cultural Policy," *International Journal of Cultural Policies* 7, no. 1 (2004): 110.
6. Cunningham, "The Creative Industries," 110–11.
7. Cunningham, "The Creative Industries," 108–9.
8. Elisabeth C. Economy, *The Third Revolution: Xi Jinping and the New Chinese State* (New York: Oxford University Press, 2018), 123.
9. Yiwei Wang, *China Connects the World: What behind the Belt and Road Initiative* (China Intercontinental Press, 2017).
10. See Jia and Nieborg in this volume; see Lin and de Kloet in this volume; Stuart Cunningham, David Craig, and Junyi Lv, "China's Livestreaming Industry: Platforms, Politics, and Precarity," *International Journal of Cultural Studies* (2019): 2.
11. Luciana Lazzeretti, Francesco Caponea, and Erdem Seçilmiş, "In Search of a Mediterranean Creativity: Cultural and Creative Industries in Italy, Spain and Turkey," *European Planning Studies* 24, no. 3 (2016): 582.

12. Lazzeretti, Caponea, and Seçilmiş, "In Search of a Mediterranean Creativity," 583.
13. Erdem Seçilmiş and Mehmet C. Güran, *Ankara Kültür Ekonomisi: Sektörel Büyüklüklerin Değerlendirilmesi* (Ankara, Turkey: T.C. Kültür ve Turizm Bakanlığı, 2013), 65.
14. Mutlu Binark and Günseli Bayraktutan-Sütçü, *Kültür Endüstrisi Ürünü Olarak Dijital Oyun* (Istanbul: Kalkedon, 2008).
15. Mutlu Binark and Günseli Bayraktutan-Sütçü, "A Critical Interpretation of a New 'Creative Industry' in Turkey: Game Studios and the Production of Value Chain," in *Computer Games and New Media Cultures: A Handbook of Digital Games Studies*, ed. Johannes Fromme and Alexander Unger (Heidelberg, Germany: Springer, 2012).
16. The Ministry of Development, *The Information Society Strategy Renewal Project: Information Technologies Sector Current Status Report* (Ankara, Turkey: Ministry of Development, 2013), 14.
17. The Ministry of Development, *The Information Society*, 88.
18. The Ministry of Development, *The Information Society*, 94.
19. The Ministry of Development, *The Information Society*, 106.
20. The Ministry of Development, *The Information Society*, 142.
21. Ministry of Development Department of Information Society, *The Information Society Strategy and Action Plan (2015–2018)* (Ankara, Turkey: Ministry of Development, 2015), 40–65.
22. Ministry of Development Department of Information Society, *The Information Society*, 76–87.
23. Ministry of Development Department of Information Society, *The Information Society*, 87.
24. Haşmet Gürçay, Emek B. Kepenek, and Engin C. Tekin, *Türkiye'de Dijital Oyun ve Animasyon* (Ankara, Turkey: Retro, 2019), 55.
25. Gürçay, Kepenek, and Tekin, *Türkiye'de Dijital Oyun*, 55.
26. Serra Sezgin, "Digital Games Industry and Game Developers in Turkey: Problems and Possibilities," *Moment Journal 5*, no. 2 (2018): 243–44, https://doi.org/10.17572/mj2018.2.238254.
27. Gürçay, Kepenek, and Tekin, *Türkiye'de Dijital Oyun*, 52.
28. Gürçay, Kepenek, and Tekin, *Türkiye'de Dijital Oyun*, 48–50.
29. Online Games Worldwide, February 1, 2020, https://www.statista.com/outlook/212/100/online-games/worldwide.
30. Max Haiven, *Hayali Sermaye*, trans. Yasin E. Kara (Istanbul: Koç Üniversitesi Yayınları, 2016), 143.
31. Haiven, *Hayali Sermaye*, 145.

32. Hye-Kyung Lee, "The Political Economy of 'Creative Industries,'" *Media, Culture & Society* 39, no. 7 (2017): 9.
33. Alex Means, "Biyopolitik Ekonomide Bir Eğitim Sorunsalı Olarak Yaratıcılık," in *Bilişsel Kapitalizm*, ed. Michael A. Peters and Ergin Bulut (Ankara, Turkey: Notabene, 2014), 299.
34. Means, "Biyopolitik Ekonomide Eğitim," 311.
35. John Hartley et al., *Key Concepts in Creative Industries* (London: Sage, 2013), 112.
36. Aaron Smuts, "Are Videogames Art?," *Contemporary Aesthetics* 3 (2005), https://contempaesthetics.org/newvolume/pages/article.php?articleID=299.
37. See Mia Tokumitsu, *Do What You Love and Other Lies about Success and Happiness* (New York: Regan Arts, 2015) and Ergin Bulut, "Playboring in the Tester Pit: The Convergence of Precarity and the Degradation of Fun in Video Game Testing," *Television & New Media* 16, no. 3 (2015) for a broader and deeper perspective to the subject.
38. Mia Tokumitsu, *Do What You Love and Other Lies about Success and Happiness* (New York: Regan Arts, 2015), 2.
39. Ergin Bulut, "Playboring in the Tester Pit: The Convergence of Precarity and the Degradation of Fun in Video Game Testing," *Television & New Media* 16, no. 3 (2015): 246, https://doi.org/10.1177/1527476414525241.
40. Sezgin, "Digital Games Industry," 250.
41. Ceren Mert-Travlos, "The Duality of Creative Hubs in Non-Western Contexts: The Case of Bomontiada," *Cultural Trends* 5, no. 2 (2021): 4.
42. Serhan Ada, "For a New Cultural Policy," in *Introduction to Cultural Policy in Turkey*, ed. Serhan Ada and Ayça İnce (Istanbul: Istanbul Bilgi University Press, 2009), 93–94.
43. Jean-François Polo, "Turkish Cultural Policy: In Search of a New Model?," in *Turkish Cultural Policies in a Global World*, ed. Muriel Girard, Jean-François Polo, and Clemence Scalbert-Yücel (Cham, Switzerland: Palgrave Macmillan, 2018), 78.
44. Mutlu Binark and Günseli Bayraktutan-Sütçü, *Kültür Endüstrisi Ürünü Olarak Dijital Oyun* (Istanbul: Kalkedon, 2008).
45. Asu Aksoy, "The Atatürk Cultural Center and JDP's 'Mind Shift' Policy," in *Introduction to Cultural Policy in Turkey*, ed. Serhan Ada and Ayça İnce (Istanbul: Istanbul Bilgi University Press, 2009), 193.
46. Mutlu Binark and Günseli Bayraktutan, "Discursive Strategies and Political Hegemony in Turkish Politics: The Justice and Development Party's #yedirmeyeceğiz vs #occupygezi," in *Authoritarian and Populist Influences in the New Media*, ed. Sai Felicia Krishna-Hensel (London: Routledge, 2018), 9–38.

47. For further information see Binark and Bayraktutan, "Discursive Strategies and Political Hegemony."
48. Hakan Yavuz, "Social and Intellectual Origins of Neo-Ottomanism: Searching for a Post-National Vision," *Die Welt Des Islams* 56, no. 3–4 (2016): 439, https://doi.org/10.1163/15700607-05634p08.
49. Yavuz, "Social and Intellectual Origins," 433–77.
50. *Fetih 1453*, directed by Faruk Aksoy (2012), is a movie that narrates the conquest of Constantinople by Fatih Sultan Mehmet and its transition from an Orthodox Christian city into the Sunni-Muslim city of Istanbul. *Diriliş Ertuğrul* is a historical television serial, produced by TRT (Turkish Radio TV Institution), that narrates the empire's early years. TRT promotes the series worldwide as a tool of soft power.
51. In similar manner to Erdoğan, innovation is central to Xi Jinping's notion of rejuvenation of the great Chinese nation (Elizabeth Economy, *The Third Revolution: Xi Jinping and the New Chinese State* [Oxford: Oxford University Press, 2018], 134). But Xi is not the first Chinese leader to emphasize the importance of innovation for China's economic rise. According to Economy, "Former Chinese president Hu Jintao also delivered a lengthy speech in 2007, calling innovation 'the core of our national development strategy and a crucial link in enhancing the overall national strength. Innovation is a matter of both economic imperative and national pride" (*Third Revolution*, 135).
52. Arjun Appadurai, "Demokrasi Yorgunluğu," in *Büyük Gerileme*, ed. Heinrich Geiselberger (Istanbul: Metis, 2017), 19.
53. Appadurai, "Demokrasi Yorgunluğu," 21.
54. See Türk, "Muktedir"; Binark and Bayraktutan, "Discursive Strategies and Political Hegemony."
55. The speeches are accessed from the Presidency of the Turkish Republic's website, December 1, 2019, https://www.tccb.gov.tr/en/receptayyiperdogan/speeches/.
56. May 8, 2018.
57. The first National Culture Congress was convened in 1982, and the second one was held in 1989. The third one took place during 2017 to develop and enrich national culture and produce a new cultural policy in line with contemporary requirements. The reports, published by the Ministry of Culture and Tourism, were prepared by seventeen subcommissions. Although these reports are on different topics, they commonly recognized information technologies, the internet, social media, and digital games as a policy field.
58. Erdoğan's speech to the Third National Culture Congress on March 3, 2017.

59. In religious rhetoric, "creation" means all of the features of a human being that have not been subject to any outside effect and that a person has at birth. Things like soul cleansing and faith in God come to the forefront in this particular interpretation.
60. For additional information, see https://www.iletisim.gov.tr/English/haberler/detay/erdogan-we-must-stand-on-our-own-feet-in-all-areas-from-data-production-to-data-security-from-defense-health-and-it-to-ai.
61. See the Digital Transformation Office's Twitter account (@dijital): https://twitter.com/dijital.
62. For additional information see https://cbddo.gov.tr/haberler/4259/cumhurbaskani-erdogan-hgm-atlas-ve-hgm-kure-uygulamalarinin-tanitilmasi-toreni-nde-dijital-gelisimi-vurguladi.
63. For more information, see https://nusrat.gsb.gov.tr/.
64. Of the total number of those dismissed by the decrees (125,806), 6,081 are academicians. See https://bianet.org/bianet/siyaset/198965-701-sayili-khk-ile-18-baris-akademisyeni-ihrac-edildi.
65. Mariano Martín Zamorano, "Reframing Cultural Diplomacy: The Instrumentalization of Culture under the Soft Power Theory," *Culture Unbound: Journal of Current Cultural Research* 8 (2016): 178.
66. Polo, "Turkish Cultural Policy," 95.
67. Economy, *The Third Revolution*, 141.
68. Economy, *The Third Revolution*, 141.
69. The Ministry of Industry and Technology and the Development Agencies established in different regions of the country have start-up support, investment incentives, marketing, and promotion grants, as well as tax reductions especially for those companies founded within the techno-cities.
70. For additional information, see https://www.tccb.gov.tr/haberler/410/114129/-yeni-teknolojileri-ureten-tum-dunyaya-yayan-bir-ulke-h-line-gelmekte-kararliyiz-.

Making, New Shanzhai, and Countercultural Values

Ethnographies of Contemporary, Innovative, and Entrepreneurial Digital Fabrication Communities in Shenzhen, China

Daniel H. Mutibwa and Bingqing Xia

It is approaching ten years since *making* and the Maker Movement are said to have reached China. During that time, *making* has been framed variously and associated with aspirations and values emerging from the international countercultural struggles spanning the late 1950s to the late 1970s. In this chapter, we explore the extent to which the framings of *making* are discernible in, and reflective of, countercultural values in the Chinese context—with a particular focus on the city of Shenzhen. To guide our exploration, we ask:

1. What does making in Shenzhen reveal about the identities and composition of its digital fabrication communities?
2. In which ways do the aspirations and motivations of these communities reflect countercultural values?
3. Where countercultural values are discernible, how are they reconciled with entrepreneurial motivations and institutional agendas in an effort to achieve change (broadly defined)?

In response to these questions, we organize the chapter as follows. We discuss how *making* and countercultural values have been framed globally and in the Chinese digital fabrication scene. We then introduce the ethnographic methodology and associated approaches underpinning the research we report on. Thereafter, we discuss the identities and composition of selected Maker communities and explore the extent to which their aspirations and motivations harmonize (or not) with countercultural values, entrepreneurial motivations, and institutional agendas in the quest to achieve change.

Making reference to carefully selected and seemingly interconnected socioeconomic and innovation-led developments in the United States and China, we argue that *making* in Shenzhen does exist as a significant grassroots countercultural current that offers Maker communities a new era of unparalleled autonomy to pursue opportunities for commons-based peer production on their own terms, and to help improve society—albeit with serious caveats. One caveat is that this autonomy is made possible by the Chinese state. Other caveats constitute entrepreneurial motivations and institutional agendas that pose major challenges. One challenge is a process we call "inversion" whereby state agendas (mis)appropriate countercultural values for their own ends, thereby undermining what such values strive to achieve. This process of inversion institutionalizes and instrumentalizes *making*, thereby rendering it a productive activity that is to be understood as a contained experiment and instrument—supported and co-opted to contribute to China's national modernization project and global economic and political ambitions.

Conceptualizing *Making* and Countercultural Values

The term *making* encompasses a number of different meanings. The most common one alludes to the creation, building, design, and modification of—or tinkering with—just about anything ranging from physical objects such as traditional crafts created manually using wood, clay, or plastic to "digital fabrication."[1] The latter refers to processes that use computer-controlled tools to make materials. This is the sense in which we use and mean *making* throughout this chapter. Behind *making* is the Maker Movement, which is said to have originated in the United States. The Movement encourages people to access, learn, and utilize digital fabrication to experiment, play, and create things informed by their imaginativeness and innovation. The understanding is that not only do people get to shape technology on their own terms, but they also get the opportunity to contribute to making their world a better place.[2]

Widely seen to provide access to tools and resources to experiment, learn, repurpose, create, and make unique products and services through hands-on involvement either in self-directed or collaborative design and digital fabrication projects, *making* in associated makerspaces and fabrication laboratories (commonly abbreviated as Fab Labs) has been broadly framed in five interconnected ways. First, *making* signals a "'third industrial revolution' extending the digital revolution of ICTs and social media into the material world."[3] Second, it champions a "'democratisation of manufacturing' as citizen-consumers engage in commons-based peer production." Third, it unlocks "'grassroots innovation' and entrepreneurship through accessible digital fabrication." Fourth, it facilitates "more 'sustainable production and consumption' through local provision, remanufacture, and the material cultivation of post-consumerist values."[4] Lastly, it provides possibilities for hands-on learning that is particularly aimed at engaging young people in science and engineering.[5] In addition, *making* has been reported to facilitate the development of other core skills such as "problem-solving, risk taking, creativity and the confidence to experiment."[6] These skills draw on a synthesis of approaches from science, technology, engineering, arts, and mathematics (STEAM) subjects in ways that conventional education may not be able to.[7]

The first two modes of framing speak to a "web innovation model" that has decentralized technologies, lowered the barriers to entry into production infrastructure and supply chains,[8] revolutionized manufacturing,[9] and challenged the status quo by way of subverting perceived dominant copyright regimes[10] and conventional processes of organizing, innovating, and engaging with material and mental resources, among other aspects. The third form of framing points to opportunities to enable personalized design and digital fabrication not only for artistic and personal ends, but also for entrepreneurial purposes.[11] The last two framing modes highlight the ability to provide solutions for sustainability and to nurture alternative ways of producing, acquiring, and sharing knowledge.[12] Framing *making* in these ways is reminiscent of some prominent analyses of the international countercultural struggles between the late 1950s and the late 1970s that have explored the central aspirations and values underpinning those struggles.[13]

Recent scholarship has associated *making* with movements, entities, and individuals in the said struggles that championed deviation from dominant norms, views, and conformity.[14] These occurrences have been perceived differently.[15] In this chapter, we take the view that countercultural activity broadly constituted a radical break with existing ideals and norms in search of personal and collective freedom. Countercultural activists viewed themselves as operating either against or outside the confines of the status quo. This manifested itself in the adaptation of prevailing ways of being and communication to achieve change (broadly defined), and in the adoption of perceived alternative lifestyles and practices as means of expressing dissent. Subversion of received wisdom and related ideals, coupled with experimentation with ideas, different forms of expression and being, and just simply creating and repairing things, were integral elements of countercultural activity.[16] All this has been seen to convey "an optimistic narrative of partially thwarted social progress that nonetheless could be completed one day in the future [through] 'self-cultivation, self-direction, self-understanding, and creativity.'"[17]

This reading aligns with *making*. Scholars have observed the ways in which *making* is strongly reminiscent of "countercultural protest" and exudes "counter-cultural roots."[18] In this narrative, countercultural activity and *making* can be said to deplore conformity, to distrust routine, to encourage resistance to established power, and to challenge standardized ways of knowing, seeing, and thinking. Both are impelled by a sense of injustice,[19] share a disillusionment with the agendas imposed by the state and capitalist excesses, and show a disdain for dominant forms of organization—be they of a social, political, economic, policy, or technological nature.[20] Countercultural activists and makers are largely white, middle-class, relatively young, usually well-educated people (barring exceptions) who have looked beyond material factors alone in search of deeper meaning in their lives. That meaning—which is thought to derive in part from frustrations of unfulfilled expectations and disaffection with scarce opportunities for empowerment and fulfillment—can be said to comprise the desire to achieve unrealized spiritual needs and ideological orientations that the said constituencies believe are being stymied by powerful capitalist forces, standard etiquette, and structural inequalities.[21]

This conceptualization operationalizes countercultural values and *making* as embodying acts of protest following Dieter Rucht's useful characterization of protest. In this view, countercultural activists and makers are bound together by a fivefold aspiration: to give voice to concerns, problems, or critiques that speak to local, regional, national, and international matters; to express and/or communicate a minority or majority position; to refer to material or immaterial goods; to envisage a short- or a long-term perspective; and to aim at a minor political change or a fundamentally different societal order.[22] In the United States, these features have demonstrably characterized grassroots countercultural activity in its pursuit of change. By contrast, efforts geared toward change in China have been consistently engineered by the Chinese Communist Party (CCP) via a top-down approach that taps into some characteristics of protest to accelerate China's nationalist project to modernize without undermining the standing of the state bureaucracy.

It is worth noting that the United States and China have co-opted the countercultural rhetoric of disruptiveness, pragmatism, experimentation, and broader economic prosperity to try and foster socioeconomic and innovation-led development. We call this the process of "inversion" whereby institutional agendas in both countries have (mis)appropriated countercultural values for their own ends, thereby undermining what such values intend to achieve. Nowhere has this process been more prominent than in the cities of Las Vegas and Shenzhen respectively—both of which enjoy exceptional global symbolism. In both cities, authorities have been an "absent presence"—allowing informality, risk, and experimentation to drive socioeconomic and innovation-led development for nearly four decades.[23] This is one example where the innovation-driven paths to development in the United States and China appear to be connected. We focus on Shenzhen because of its extraordinary status as a Special Economic Zone and a global model hardware and IoT city—thanks to the machinations of the CCP over time.

Making, *New Shanzhai*, and Countercultural Values: Digital Fabrication in the Chinese Context

In the ethnographic inquiry we report on in this chapter, we explored the extent to which the framings of *making* outlined earlier are discernible in the Chinese Maker context. Commentators have noted that China's status as the manufacturing capital of the world—reflected in the ubiquity of the "Made in China" label—over nearly forty years developed an open, low-cost production, network model that mirrored some overriding principles of the Maker Movement and related countercultural values.[24] One such principle is the understanding that solutions to key societal challenges can be searched for and arrived at differently.[25] As already hinted at—and as we shall see—the Chinese state has co-opted these principles for its own ends.

During the 2000s, the aforementioned model coevolved with *Shanzhai*[26]—which is understood to mean "copycat" goods or "counterfeit" products. With the emergence of Maker culture in China around 2010, a convergence of the Maker Movement and *Shanzhai* has culminated in what David Li—seen widely by many in China and abroad as one of the pioneering leaders of the Chinese Maker

Movement—has termed *New Shanzhai*.²⁷ *New Shanzhai* is perceived to tap into free, informal, and open-source systems and infrastructures involving hardware, electronics, new forms of manufacturing, and information-sharing.²⁸ In Shenzhen, the *New Shanzhai* ecosystem is characterized by four key factors—namely, (a) speed of innovation, (b) a disregard for intellectual property (IP) protection, (c) a dense network of communities of makers, entrepreneurs, venture capitalists, businesspeople, hardware start-ups, accelerators or incubators, innovation hubs and centers, various intermediaries and other stakeholders, and (d) institutional support.

The need to respond to niche market demands not catered to by big businesses has significantly boosted the ability to prototype fast and assemble components in different formats swiftly, something that demonstrates an almost instant provision of innovative solutions.²⁹ A key feature of this approach is that "weaker IP protection [where it exists] and cut-throat competition mean that [makers] and entrepreneurs place less emphasis on protecting their inventions in China, instead attempting to innovate quicker than their competitors."³⁰ Key competitive advantages are a familiarity with local tastes both at home and abroad, as well as the proximity to the world's fastest growing markets in India and Southeast Asia.³¹ A further crucial competitive advantage is the ability to tap into large networks of stakeholders embracing an open, experimental manufacturing culture³² increasingly supported by policy initiatives and subsidy schemes in line with the ambition to foster innovation-led economic development geared toward a "Designed in China" status.³³

Of particular note is the "Made in China 2025" initiative—commonly abbreviated as "MiC2025." Since 2013, MiC2025 has been facilitating the transition from the aforementioned low-cost production, network model to "intelligent" or "smart" manufacturing, which is understood as the incorporation of communication technologies and advanced digitally mediated infrastructure into various stages of manufacturing processes across design, production, branding, and delivery of products and services.³⁴ This mode of manufacturing has gradually been introduced in a range of industrial sectors pertinent to *making* such as electronics, telecommunications, household appliances, and advanced internet services, among many others.³⁵

These developments have significant implications. First, grassroots innovation, design, and digital fabrication in China is increasingly being (mis)understood as inherently business-oriented—a criterion used to allocate public subsidy.³⁶

Second, Maker communities envision the future of innovation differently from government.[37] Some commentators have reported on the belief widely held by the said communities that only individual and collective empowerment—in tandem with appropriate support for bottom-up approaches—will foster effective and sustainable social and economic transformation.[38] Third, the reliance on funding from government, institutions, and venture capitalists raises the question whether top-down support can potentially clash with the ethos of Maker culture.[39] Indeed, a central critique of "MiC2025" has been that its top-down orientation has not only stultified organic and bottom-up dynamics in some instances, but it has failed to coordinate policies at central and state and/or local levels more effectively.

Fourth, a transition to innovation-led economic development requires free expressions of creativity and thought that the traditional Chinese education system premised on rote learning is currently unable to provide.[40] It is widely acknowledged that increased investment in *making* and the learning of related competencies and skills via Maker Education (also abbreviated as Maker Ed) is essential in supporting the growth and effectiveness of grassroots creativity and innovation. Fifth, the state of flux of grassroots innovation and digital fabrication "has produced manic and fierce competition among swarms of entrepreneurs."[41] This state of affairs—as it plays out in Shenzhen—reflects the paradoxical nature of the "*change/continuity* and *disruption/structure*" dyads that the editors of this volume outline in the introduction. In the next section, we discuss the methodology we employed, the case studies we analyzed, and the ethics-related considerations we abided by.

Method and Case Studies

In this chapter, we explore how Maker communities in Shenzhen "engage materially with digital fabrication, and how observed practices shape, enable, and underpin the formation, validation, or unsettling of [the framing of *making* and countercultural values outlined earlier]."[42] This exploration draws on a comparative case study approach[43] and strategies from the co-production research tradition,[44] which allow research participants to provide as well-rounded a picture as possible of their world. We situated the said approach and tradition within an overarching ethnographic methodology[45] that allowed us to draw on fifty-one semi-structured qualitative interviews, participant observation, the study of documentary evidence, and six focus groups in the examination of the communities of makers, entrepreneurs, hardware start-ups, and other stakeholders that we were able to access between March 2017 and April 2018.

For our purposes in this chapter, we have selected four sites that are reflective—but not necessarily representative—of makerspaces and hardware innovation hubs (entrepreneurial and non-entrepreneurial alike) in Shenzhen. These selection

TABLE 1. Case Study Sites and Interviewee Information

MAKER SITES	PARENT ORGANIZATION(S)	INTERVIEWEES[*]
SEGMaker[†]	Shenzhen Electronics Group (SEG)[‡]	Yuong Jay[§]
SZOIL[¶]	Maker Collider[#] and Shenzhen Industrial Design Profession Association (SIDA)[**]	Vicky Xie[††]
Chaihuo x.factory[‡‡]	Seeed Studio[§§]	Violet Su[¶¶]
Litchee Lab[##]	N/A	Lit Liao[***]

Notes:
[*] The interviews lasted between 60 and 90 minutes on average. In accordance with the ethical terms under which access to conduct fieldwork was granted, we reveal the true identities of our interviewees including their real names.
[†] SEGMaker comprises a co-working space and an incubator. It provides tailored services in the form of capital, equipment, training, and access to various Maker and hardware communities with the ultimate goal of nurturing hardware start-up entrepreneurship. SEGMaker is part of the Fab Lab network—a community of fabricators and designated community-based spaces located in over 100 countries across the globe. For more information, visit https://fabfoundation.org/getting-started/#fablabs-full and https://www.fablabs.io/labs/segmaker.
[‡] SEG is a local government institution which—for over thirty years—has been operating in the electronics product-development business sector among others.
[§] Yuong Jay is Operations Manager at SEGMaker. He holds a postgraduate degree in project management and has corresponding work experience—including basic knowledge of hardware.
[¶] SZOIL is a unique hub that comprises an open innovation center and a start-up accelerator. It views itself as "a space and platform for worldwide makers to communicate and cooperate," and as a "global maker service platform" and promoter of Shenzhen as a city hub of "digital intelligent hardware and manufacturing." In this capacity—and as part of the Fab Lab network, it undertakes R&D activities, offers innovation and entrepreneurial training and education courses for makers, and provides a bespoke "industry chain collaboration service." For more information, access http://www.szida.org/list-31-1.html.
[#] Maker Collider is an open-source Maker platform sponsored by Intel China. For further details, visit http://www.makercollider.com.
[**] SIDA is a local government-funded, nonprofit organization that connects industrial design and makers (both local Chinese and foreign) across the globe. See also http://www.szida.org/list-31-1.html.
[††] Vicky Xie is the Global Corporation Director at SZOIL. She holds a bachelor's degree in English language studies.
[‡‡] Chaihuo x.factory describes itself as an "open factory" furnished with "production-level equipment for in-house prototyping and small-batch production services as well as co-working spaces" among other offerings. It was established in 2011 as Chaihuo Maker Space—and is widely considered to be a pioneering makerspace in Shenzhen. Chaihuo x.factory is a subsidiary entity of Seeed Studio. See https://www.seeedstudio.com/about_seeed.
[§§] Seeed Studio describes itself as an Internet of Things (IoT) hardware enabler that has been providing open-source hardware and manufacturing services to global makers and other stakeholders since 2008. See also https://www.seeedstudio.com/about_seeed; https://www.seeedstudio.com/blog/; https://twitter.com/seeedstudio.
[¶¶] Violet Su is an English language studies university graduate and the community manager in charge of programs and projects at Chaihuo x.factory.
[##] Litchee Lab is a makerspace that offers bespoke Maker Ed curriculum services to schools and other learners while also providing access to a space and equipment for personal and collective fabrication. It is part of the Fab Lab network. See http://www.litchee.cn/.
[***] Lit Liao is the founder of Litchee Lab. She is a university graduate of electronic engineering and one of the movers and shakers of Maker Ed in Shenzhen.

criteria are supported by the status and reputation these sites enjoy in key Maker circles in the region.[46] The four sites are SEGMaker, Shenzhen Open Innovation Lab (SZOIL), Chaihuo x.factory, and Litchee Lab. Information about the sites and corresponding research participants is presented in tabular form. Details pertaining to background and context are presented in footnotes.

Shenzhen Maker Communities in Action

In what follows, we explore who the Maker communities at the selected case-study sites in Shenzhen are; what their aspirations and motivations are; how these reflect countercultural values or not; and whether or not those values are reconcilable with entrepreneurial motivations and institutional agendas in the quest to achieve change (broadly defined).

Identities and Composition of Maker Communities in Shenzhen

With respect to the identities and composition of the Maker communities under study here, our interviewees described the majority of the makers and hardware businesses as constituting both local and other inland Chinese nationals. Yuong Jay helpfully quantifies this effectively. Of the "over 200 [digital fabrication] projects" at SEGMaker, 180 or so are owned by the said Chinese demographics. Since the same can be said of the other three sites, this could be said to reflect the exponential uptake of grassroots design, innovation, and *making* that is heavily supported by the mass entrepreneurship and innovation scheme. This scheme is an integral component of "MiC2025" and is tasked with tapping into citizens' everyday creativity.[47] However, the prevalence of international makers is striking. This "very internationalized" Maker scene, as Violet Su noted, renders *making* in Shenzhen quite unique due to the city's distinctive *New Shanzhai* ecosystem that attracts makers and other hardware stakeholders from inland and Greater China, numerous parts of East and South Asia, and across the globe.

For these local and international constituents, "Shenzhen is the optimal place to develop hardware, both in terms of speed to market and efficiency working the supply chain."[48] It is not surprising, then, that these aspects—and other draws—have led to Shenzhen being (informally) called, and globally celebrated as, "the Mecca of Hardware" among many other labels.[49] Apart from the international dimension, our interviewees described the makers at their sites as varied in terms of background. There was mention of "hobbyists," "hardware entrepreneurs," "start-ups," "university students," and "artists." We learned that a number of individuals and Maker teams worked as technicians of various kinds for some of the Chinese companies and factories that manufacture products for the big global high-tech brands. Overall, individuals tended to have what Yuong Jay refers to as "common knowledge" understood as either a background in science and engineering or a familiarity with design and digital fabrication processes, albeit to differing levels.

With the exception of SEGMaker, which appears to be fairly exclusive in its support for male makers with bachelor's degrees ideally in the STEAM subjects, the other three sites come across as fairly inclusive all around. This inclusivity is clearly reflected in the descriptions of our interviewees in the preceding section. There, we see that a university degree qualification is common. This appears to be a prerequisite to belong to the respective sites in an administrative and/or management capacity, but is not necessary for an ordinary site member, or a member of the wider Shenzhen Maker scene. Similarly, we see that *making* is very much open to females as it is to males, and that belonging is not dependent on having "common knowledge"—though bringing along an openness and willingness to learn digital fabrication skills and a passion for *making* is essential, be it for site administration and/or management purposes or for self-actualization, or both.

Although not mentioned explicitly in the accounts provided by our interviewees, we noted as participant observers that the demographics of the makers under discussion—local Chinese and foreign alike—seemed overwhelmingly young. That is to say, individuals generally appeared to be either below or around the age of thirty years. This chimes with known accounts that have recorded the average age of Shenzhen residents as twenty-seven years old or marginally higher.[50] Having established the identity and composition of the Maker communities under discussion here, we now look at what drives them.

Aspirations and Motivations: (Mis)alignment with Countercultural Values?

Earlier on in the chapter, we conceptualized *making* as it aligns with countercultural values. We now turn to our interviewee accounts that capture the aspirations and motivations that impel Maker communities in Shenzhen in an attempt to make sense of how such aspirations and motivations reflect countercultural values or not:

> So, in the weekend we'll have workshops in this area for the primary school students or senior school students for them to really experience a "fantasy [land]" of hardware. So, the workshops teach them how to build their own hardware, how to build their own cars, their own drones, so things like that. (Yuong Jay, SEGMaker)

> [We tend] to say: "everyone is a maker but not everyone is an entrepreneur." ... We are supporting different groups [including] hobby[ists], hardware start-up teams or entrepreneurs. They're actually having a good environment now—government support, makerspaces, accelerators, factories, manufacturers, industrial design companies, independent design houses.[51] They've got all the best resources available here in Shenzhen. (Vicky Xie, SZOIL)

> So, all these resources ... help our makers to grow. And our makers can also provide some solutions to the industries [as well as to problems in everyday life]. So, the open-source [ecosystem] help[s] [makers] to lower the cost and also help[s] start-ups to get into [hardware business]. So, there are very versatile and very different kind[s] of projects, no matter it's for fun or for solving a problem or adding some colors to our life. (Violet Su, Chaihuo x.factory)

> [A] challenge that China is facing now [is] how to prepare our next generation[s] for [the] future and Maker Ed is doing that by teaching our kids how [to use] the "learn and work model."[52] (Lit Liao, Litchee Lab)

These insights relate to aspirations and motivations at two levels, namely, the case study sites and the Maker communities they claim to serve. The sites play what appears to be an intermediary role that involves nurturing and bringing Maker communities into contact with each other and with a range of other actors and resources both within the *New Shanzhai* ecosystem and across the entire supply chain. Nurturing involves offering training and learning as well as the provision of opportunities and resources for Maker communities to *make* things that excite them and are meaningful to them.[53] We have seen that one way to achieve this is to get some of these communities "to really experience a 'fantasy [land]' of hardware"—whether or not the things made are seen to be "useless." Regardless of whether or not some of the stakeholders are motivated by entrepreneurial objectives, the sites link them to "government support," design-service providers, personalized manufacturing services at low cost, educational entities and associated curricular networks, and "marketing and sales" contacts, among other things.

In doing so, the sites can be said to structure, sequence, and pace digital fabrication projects in Shenzhen in a process that is characterized by a fourfold dimension: proselytizing, matchmaking, gatekeeping, and counseling.[54] This mode of operation originated in Silicon Valley in California in the United States, and has been gradually transposed to Shenzhen through industrial flows (know-how and labor) in which the big high-tech brands have been subcontracting Shenzhen industrial partners to manufacture cheaply for them. This is another example of how creativity and innovation in China appears to be inextricably linked to innovation-driven developments in the United States at different junctures in time. By undertaking these overarching functions, the sites under analysis here have "becom[e] cradles of entrepreneurship, innovators in education, nodes in open hardware networks, studios for digital artistry, ciphers of social change, prototyping shops for manufacturers... emblematic anticipations of commons-based, peer-produced post-capitalism... galleries for hands-on explorations in material culture... and not forgetting, of course, spaces for simply having fun."[55]

At the level of Maker communities, individuals appear to be driven by the desire to address gaps and needs in provision prevailing in society. Violet Su alludes to makers who have the potential to "provide some solutions" to problems that may be perceived to have been ignored by established actors in the areas of economics, politics, technology, and policy, among others. The desire, aspirations, and motivations to make the world a better place have been associated with shared imaginaries and activism that embody the position of being in a struggle for critical, egalitarian (and sometimes alternative) ways of being, seeing, thinking, and acting.[56] Some of these aspects manifest themselves well in Lit Liao's vision of the potential of Maker Ed. Its intervention is premised not only on instilling in learners constructivist approaches to learning and learning by doing that challenge the conventional Chinese rote education system, but also on the belief in the transformative power of the said approaches to change many realms of society. One realm that Lit Liao makes particular reference to is "smart" manufacturing—a sector she believes China can excel and become a world leader in in the foreseeable future. We place this later in the broader context of China's ambition

FIGURE 1. 3D printing room, Chaihuo x.Factory (previously known as Chaihuo Maker Space)

to redeploy processes of transformation to secure national and global economic and reputational advantages.

The increasing proliferation of Maker Ed projects in Shenzhen—and the subsequent public debates they have triggered in relation to the competencies and skills needed to facilitate innovation-led economic development—appears to support a key account. When people like Lit Liao mobilize resources in accord with (critical) ideas, they have the possibility of contributing to the restructuring of established formalities and systems.[57] In these and other ways, the aspirations and motivations of some of the case study sites and their Maker communities reflect not only countercultural values, but also elements of protest as conceptualized earlier.

From a collective vantage point, these sites encourage grassroots design, innovation, digital fabrication, and broader participation in industrial production processes. The sites achieve this through organizing horizontally and flexibly—in the same way that firms, start-ups, and innovation hubs in Silicon Valley have tended to organize. This allows for the employment of agile and efficient methods of working to accommodate low-cost, small-scale production that lowers the barriers to entry and dispenses with centralized, industrial-scale production typically controlled by a smaller number of hierarchically organized, established individuals and entities.[58] This clearly represents another way of organizing, one that deviates from conformity and attempts to truly democratize innovation and manufacturing processes. In doing so, the sites attempt to subvert control and domination of the global high-tech brands. But as we shall see below, the sites—and entrepreneurial Maker communities—are partly implicated in the very same modes of control and domination.

From a personal perspective, we argue that the Maker communities under study in this chapter are drawn to *making* as a lifestyle because "it helps [them] better themselves, provides both pleasure and useful skills, and ultimately frees [them] to take control of [their] li[ves]."[59] They utilize the infrastructure, training, and resources made available to them to access shared material and immaterial goods normally inaccessible outside of designated digital fabrication spaces, to experiment and learn by doing in collaboration with like-minded peers, and to become part of a community characterized by a shared ethos. An integral part of that ethos is informed by the desire to effect social change through being involved in communal attempts to improve societal life in the same way that countercultural activity does. Particularly in a cultural and educational context where following many rules, saving face, keeping valuable information to oneself, and the display of relatively little tolerance of failure and mistakes are said to be the norm,[60] the aspirations and motivations expressed above indeed appear to reflect a refreshingly new, grassroots countercultural current.[61]

To borrow David Gauntlett's (2018) words, this current positions making as "everyday creativity" that gives Maker communities "a sense of potency, expressive ways to connect with other people, and a sense of meaningful engagement with the world [including the chance to] exchange things, and [to] inspire each other."[62] In their different ways, then, the aspirations and motivations we have seen clearly reflect the said countercultural current in Shenzhen and parts of Mainland China in a way similar to our earlier conceptualization of countercultural values and framing of *making*.[63] We argue that the CCP bureaucracy understands the importance of these attributes in contributing toward China's modernization and has—through the process of inversion explained earlier—proactively supported their pursuit, but only as long as the state's socialist apparatus is not questioned. But can we really speak of a noteworthy grassroots countercultural current? So far, our findings point to a

FIGURE 2. Small Creativity, Big World (PHOTOGRAPH © DANIEL H. MUTIBWA, REPRINTED BY PERMISSION)

current discernible only in pockets of particular Maker communities that are not widespread either in Shenzhen or across China. A key explanation for this, we argue below, lies in the current's situatedness within a landscape characterized by conflicting ideologies, demands, and confines imposed by entrepreneurial motivations and institutional agendas.

Countercultural Values, Entrepreneurial Motivations, and Institutional Agendas

We have seen that countercultural values and *making* advocate pursuit of people's own liberation and self-direction over conformity. At a time when elements of capitalist ideology are increasingly dictating how Chinese society is being organized and how value systems and lifestyles of young people are being shaped,[64] especially in relation to *making* and its perceived economic and cultural value,[65] contradictions and tensions appear inevitable. We explore this through the lens of selected digital fabrication projects as presented by our interviewees:

> So, we have one of the projects—They're just putting a flash and also a radio on the [walking] stick, and also emergency call devices in the [walking] stick for elderly people. So, if elderly people have some accident, the emergency [services] will call. If they don't reply, it will automatically transfer to the police stations. (Yuong Jay, SEGMaker)

> We [have] just [run] an open-source village program . . . related to the beehive. [Following training, learners] started to design a beehive by themselves. So, the beehive is actually helping the bee farmers to . . . detect the [moisture], the temperature in the beehive, [and with the help of sensors to inform farmers] what time is the most suitable temperature for the bees to generate some honey. That will help those farmers to [increase] efficiency [in] honey production. (Vickie Xie, SZOIL)

> So, we have two boys who [separately] joined as members [and realized they had] so many interests in common, [something that prompted them to] build something together. . . . Then they built robots and the robots [are] not for sale or it's not that they want to start a start-up later or anything. They just have their daily job[s] and come here to build [their robots] and [have] upgraded to the new level [where the robots] can dance and to the next level that they can [be] remote-controll[ed]. So, it's like for fun. (Violet Su, Chaihuo x.factory)

In line with countercultural values and some dimensions of protest outlined earlier, the digital fabrication projects described here represent an exercise in self-direction, experimentation, independence, taking responsibility, and—to a certain extent—a search for a deeper meaning in everyday creativity, which can take either material or immaterial form.[66] To varying degrees, some of these projects exhibit a social justice dimension. In doing so, such projects display a culture that fuses personal and collective freedom and engagement with creativity and innovation for wider benefit,[67] albeit with some serious caveats. For instance, the said freedom is made possible in the first instance by the CCP for purely instrumental reasons. Through the process of inversion, the CCP seeks to drive societal transformations both at micro and macro levels. It also strives to stimulate processes of mass creativity and innovation through state-supported, top-down interventions in a bid to model a "new" modern China. This speaks to one of the central strands of this volume—"'old' futures"—presented in the introduction and about which Hoyng and Chong observe that what tends to be seen or presented as "new" may, in fact, turn out to be firmly grounded in "old power structures." Some of the decrees issued during the Cultural Revolution and many state-led interventions in post-Mao China have led to significant new societal transformations, including creative and innovative changes. Nonetheless, these have been—and continue to be—anchored in century-long sociocultural power dynamics in Chinese society.

Nevertheless, for many of the makers under discussion here, the priority is to engage with *making* for *making*'s sake—and to enjoy the benefits that come with the Maker lifestyle discussed previously while simultaneously working their day jobs or pursuing study. At the time of our fieldwork, the projects described had not been turned into entrepreneurial ventures—but that possibility remained open. Members of the project that added value to the walking stick were more interested in (*re*)*making* it for use by elderly members in their family circles. The beehive project is part of SZOIL's social program that champions self-sustenance, builds capacity, and seeks solutions to improve the living conditions of the most disadvantaged communities at the margins of the city of Shenzhen. Outside this context, the project has huge economic potential, as does the robotics project undertaken by the "two boys."

These projects—and many others like them, however—tell one side of the story. The makers of these projects are non-entrepreneurial and can be said to foreground the kind of social justice ethos that characterizes countercultural activity and some dimensions of *making* discussed earlier. These makers appear not to be interested in turning their ideas and products into entrepreneurial ventures—at least initially. Vicky Xie captures this neatly above when she remarks that "everyone is a maker but not everyone is an entrepreneur." The other side of the story is told by those makers who either attempt to balance social justice objectives and entrepreneurialism or go down the entrepreneurial route entirely (Maker entrepreneurs proper). We show how the balancing act plays out.

Makers committed to countercultural values (and by extension social justice goals) and entrepreneurialism approach *making* in the knowledge that they have access to "all the best resources available … in Shenzhen," as Vicky Xie notes above. Often, because these makers have no knowledge of business development, the sites advise them to focus on *making* and to hire services to help with business operations rather than launch into full-blown business activity. Proceeding this way helps these makers not only to concentrate on what they do best—which is to identify gaps in provision, provide solutions to common needs, or, as Violet Su noted earlier, add "some colors to our life"—but also to stay lean, agile, and flexible while working their day jobs or studying.

An example that illustrates this process—and captures the inner workings of Shenzhen's Maker ecosystem effectively—is a Maker team from Sweden that created a drone called Crazyflie.[68] Violet Su told us that this Swedish Maker team comprised three young people with day jobs who elected to balance *making* for fun with earning money from the fruits of their creativity and innovation. With no resources to progress the project from the prototype stage, the team partnered with Seeed Studio in Shenzhen, which placed the drone project on its website for preorder. Seeed Studio then invited the global hardware community to provide feedback to help continue improving the drone, but also to gauge concrete interest, which informed the company how many units of the drone were manufactured in the first batch. According to Violet Su, the preorder went viral and was even featured by *Wired* magazine.[69] In its role as an intermediary and "IoT enabler," Seeed Studio supported Crazyflie with accessing and navigating the *New Shanzhai* ecosystem and supply chain without setting foot in Shenzhen initially.

The feedback and comments obtained informed enhancements, but these had to be realized fast in order to stay ahead of both copycats and competitors since the drone was open-source. As explained earlier in the chapter, this is common practice in Shenzhen given that anyone with access to the design and product specifications can manufacture a similar product with ease. Maker entrepreneurs literally rush to get the first batch of products manufactured in order to at least recoup any investment costs, but also to try and maintain a competitive edge through improvements insofar as possible. Violet Su explained that one of the key enhancements to Crazyflie involved dispensing with the game pad used to control the drone and replacing it with an app on a smartphone instead.

At the time of our fieldwork, Crazyflie had since launched a second iteration of the product. Crazyflie is not only made merely for playful activity and entertainment, but also as a platform to build on and enhance research at affordable cost in system applications in the areas of aeronautics, robotics, logistics, and agriculture across the globe.[70] Here, Crazyflie positions itself as a Maker of a product for frivolous activity and as an entrepreneurial unit engaged in R&D. As we argue later, the latter is precisely what the Chinese state envisages the role of *making* to be in China. Institutional support in its various forms is readily made available to Maker sites that collaborate with Maker entrepreneurs like Crazyflie, either with demonstrable potential to drive design and innovation or a discernible track record in R&D. There may seem nothing countercultural about Crazyflie as a product, but there certainly is something about its ethos and way of working that reflects profoundly countercultural values encapsulated in collectivity, sharing, solidarity, breaking with convention, and leveraging innovation for truly meaningful commons-based peer production.

In balancing countercultural values and entrepreneurial motivations, Maker entrepreneurs like Crazyflie encourage open-source sharing as a means to collectively "liberate" technology from the grip of dominant high-tech companies. In doing so, they defy convention and the culture of conformity for wider benefit. However, the same cannot be said of Maker entrepreneurs proper, who are driven by solely commercial motivations. We learned that such makers tend to be less keen on solidarity and are apprehensive about sharing their ideas and projects despite utilizing open-source resources. The perceived risk of imitation and potential loss of income resulting from sharing drives such makers to sign nondisclosure agreements at the outset, and to seek to secure intellectual property (IP) protection. Although such makers have every right to proceed this way, we argue that this reproduces capitalist norms and practices such as individualism, self-interest, exploitation, and new forms of hierarchies, among others—all of which neither harmonize with countercultural values and *making* as discussed so far, nor align with social justice goals. We home in on the major implications these issues—and the developments explored earlier in this chapter—have for the wider context.

Situating *Making* and Maker Communities in Shenzhen in a Broader Context

We have seen that the identities and composition of the Maker communities at the case study sites in Shenzhen constitute overwhelmingly male, relatively young, predominantly local and inland Chinese with "common knowledge" associated either with university education in STEAM subjects or familiarity with digital fabrication processes. Females are (mostly) welcome, but they remain heavily outnumbered. One reason for this, Lit Liao and Violet Su observed, has to do with the perception that science and engineering subjects have traditionally been perceived to be male domains, while digital fabrication has tended to be associated with "geek culture."

Nonetheless, the number of female makers is steadily rising—though it will take time and a major shift in perceptions to close the gap.[71] Because digital fabrication presupposes a background in STEAM subjects and/or the mindset, competencies, and environments conducive to learning associated skills and techniques, we argue that these may be out of reach for many demographics for a range of structural reasons—including the poverty-induced illiteracy that Yuong Jay mentioned in the interview.

Far from truly democratizing design and digital fabrication, this could be seen to render *making* a minority activity, enthusiasm, and preserve of educated and other privileged individuals. This is hugely problematic because it suggests that the key actors behind—and the principal drivers and beneficiaries of—the fourth industrial revolution are likely going to be less diverse: male, young, well-educated, and technologically savvy. This threatens to undermine efforts to seek gender and class equity, to create life opportunities for all, and to liberate and leverage technology for truly broad social change. This reveals a crucial aspect that speaks to the notions of "'old' futures" and "'uncertain' futures" as explained in the introduction of this volume by the editors. Novel as the fourth industrial revolution may be, it is nestled in the usual dominant structures, which are poised to continue to dictate the contexts within which innovation occurs, and by extension, the nature of the transformations likely to be fostered. More importantly, the status of the makers described here grants them what Tyree-Hageman has termed "episteme privilege"—meaning that only they "have the means to 'exercise the causal power to undermine, dislodge, and replace a previously dominant ideational regime.'"[72]

This "regime" is central to the understanding of the relationship between culture, economics, technology, and activism across the past and present. In the United States, where this relationship has a relatively long history, key scholarly accounts have found that grassroots activists working at the intersection of culture, technology, and the economy have embraced countercultural ideals in a way that chimes with entrepreneurial success resulting from organizing and working differently than established organizations.[73] Embracing techno-libertarianism, these activists have maintained a fundamental antipathy toward state interference, which has tended to be perceived as a threat to individual freedom and civil liberties—although such activists have not seen any contradiction in obtaining government funding and other institutional support to experiment with technology projects. The success of these activists in the United States has been linked to innovative and entrepreneurial growth resulting not only from the liberation afforded by technology, but also from traits such as hard work, talent, vision, egalitarianism, transparency, and openness. Cross-pollination of ideas and practices has meant that this entrepreneurial ethic has established itself in innovation and entrepreneurial cultures, most notably in high-tech regions such as Silicon Valley in California and Route 128 in Boston.

Subsequently, some countercultural ideals have become deeply entrenched in individualist and free-market principles that characterize the high-tech industry in the United States—and now the entrepreneurial Maker communities in Shenzhen, as we have seen. This has been made possible mainly through the industrial flows explained earlier. Clearly, this has played into the hands of the "dominant ideational regime" that many grassroots activists and non-entrepreneurial makers have criticized and attempted to undermine. By embracing entrepreneurial motivations fully under the guise of nurturing the growth of society through broader civic participation, grassroots ventures seeking to make technology widely available for personal gain and commercial exploitation have implicated themselves in the sustenance and legitimation of capitalist and neoliberal approaches that have been gradually blunting, (mis)appropriating, and co-opting the said ventures. Maker entrepreneurship proper finds itself in this very same position.

Maker entrepreneurship proper in Shenzhen is characterized by the production of collective goods through the self-oriented actions of individuals and entities organized in self-interested networks that conform to market-centric economic ideals of self-reliance and wealth accumulation, among other things. In doing so, it not only invariably reproduces capitalist and neoliberal practices as we have seen, but it also selfishly exploits acts of solidarity and the open-source ethic. Moreover, the prioritization of economic interests at the expense of open-source conventions and countercultural values means that Maker entrepreneurs proper make products and services that sell to the detriment of those that address social justice issues but may be unprofitable. This epitomizes neoliberal business as usual, which renders Maker entrepreneurs proper "manifestations of flexible capitalism as much as the large IT enterprises are"—the difference being that the former do not earn as much as the latter.[74] This shows complicity in the sustenance of the "dominant ideational regime."

A crucial factor in shaping the trajectory of *making* in this way is institutional policy and associative agendas. In its quest to position itself as a promoter of innovation-led economic growth through "smart" manufacturing and as a major creator of jobs and wealth, China is heavily investing in grassroots design, innovation, and digital fabrication on a massive scale. Though similar in conception to the science and innovation programs designed to stimulate creativity en masse in the United States in the 1950s and 1960s—and latterly the "Nation of Makers" initiative in the 2010s—the distinctiveness of the Chinese mass innovation and entrepreneurship initiative lies in its coverage of far more regions and their first-, second-, and third-tier cities to stimulate economic growth on a truly monumental scale.[75] This top-down approach, which represents yet another example of how the innovation-led paths to economic development in the United States and China appear to be linked, has been criticized for selecting, institutionalizing, and turning *making* and other design and innovation activities not only into instruments of solely economic value, but also as embodiments of new, post-industrial, and urban-development identities driven by media, creative, and high-tech city agendas, among other things.[76]

For critics, the self-direction, independence, creativity, and entrepreneurial success that result from grassroots innovation and Maker entrepreneurship proper serve not only to absolve the state of its civil responsibilities, but also to help drive the CCP's national and global economic and political agendas through revision and replication of selected capitalist practices, including the mantra of the "self-made man" who makes it on his own among other perceived achievements.[77] Additionally, we learned that institutional policy agendas expect a return on investment measured in terms of the number of successful hardware start-ups, the number of products made and their value on the market, the number of jobs created, the volume of wealth generated, the rate of spillover and/or exchange of knowledge and technology within China and from abroad, and the volume of patents secured. Taking IP as an example of a return on investment, it is no coincidence that China overtook the United States in 2019 as the top nation worldwide to file patents.[78] In pursuit of these ambitions, the CCP has been said "to develop productive citizens to serve the national economy [and] to promot[e] peer production as a counter to [declining] industrial-scale economics."[79]

We argue that drafting citizens to contribute productively to the national economy in this way represents a nationalistic experiment geared toward creating a modern China in very much the same way that the agricultural and industrial decrees issued during the Cultural Revolution were nationalistic experiments designed to thrust China onto the path to modernity. Here, the change/continuity dyad that speaks to the "'old' futures' strand outlined in the introduction of this volume is discernible. Clearly, this was never intended as the raison d'être of *making*.

With entrepreneurial motivations and institutional agendas pulling *making* in Shenzhen in different directions, we return to the question of whether there exists a significant grassroots countercultural current in the city. We argue that there does exist one—albeit as an experiment that has resulted from an unusual combination of supervised institutional intervention and the state's "absent presence," something that mirrors the mode of experimentation that has characterized the socioeconomic developmental paths of Las Vegas and Shenzhen mentioned earlier. However, because of its entanglement in entrepreneurial and institutional agendas, the current is in no position whatsoever to summon up the kind of socioeconomic progress and change needed to foster truly radical transformation. Even protest and revolutionary ventures like "The Tian'anmen Incident" in 1976, "The Minzhu Qiang" in 1978/1979, and the student-led "Pro-Democracy Movement" in 1989 were far larger in scope, and yet failed to achieve radical societal change.

The current is limited in scope and is supported by the state, provided it poses no challenge to the legitimacy of the CCP. Like many spheres of society supported by institutions and capital, *making* in Shenzhen enjoys freedom and independence, but it is difficult to see how it can exceed the boundaries of autonomy acceptable to state authorities and other capital providers who support its existence. Through the process of inversion and associated institutionalization and instrumentalization, Chinese authorities have skillfully co-opted the current into the nationalist project to modernize the country. They understand that leaving Maker communities (entrepreneurial and non-entrepreneurial alike) to their own devices—which allows for greater choice, autonomy, and informality—can yield unexpected but vital productive rewards in the same way that the city projects of Las Vegas and Shenzhen have done, albeit to a varying extent.

Conclusion

The contradictions and tensions discussed above notwithstanding, *making* in Shenzhen offers a range of opportunities and possibilities around commons-based peer production that are attractive to diverse Maker communities with a variety of motivations. For some, *making* is about recreation and fun. For others, it is a source of empowerment that serves as a means to try and make society a better place. Yet for others, it is about earning income from an activity they are passionate about. For still others, it may be a combination of some or all of these things. When embraced by non-entrepreneurial Maker communities for all the possibilities it has to offer in the pursuit of change, *making* has huge potential to bring together different, like-minded individuals and groups around the task of attempting to reframe society's norms and direction.[80] *Making* can contribute to producing a different configuration—one that strives to be egalitarian and sustainable, thrives on the open-source ethos, and allows for greater possibilities. At the micro level, *making* is introducing individuals to new ways of being, thinking, seeing, and doing things. It is fostering transformative experiences as the example of the "two boys" building robots together for fun shows. Through its ethos and disruptive way of working, *making* is "connect[ing] people, ideas, regions, and countries around the world, often in unexpected ways,"[81] as the relationship between Crazyflie and Seeed Studio illustrates. *Making* is cultivating a culture of sharing, transparency, and a non-hierarchical approach to leveraging digital technology for constructivist learning, practical activity, and critical reflexivity as Lit Liao's description of Maker Ed demonstrates.

We argue that the Chinese state understands this very well, which explains why it proactively supports *making* and the pursuit of associated attributes as long as they are contained within, and do not undermine, the socialist cultural, socioeconomic, and political imaginary of the CCP. In proceeding this way, the state bureaucracy co-opts, (mis)appropriates, and instrumentalizes *making* for solely economic, political, and reputational gain, something that blunts *making* as a truly transformative societal force. Consequently, the loss of its critical, playful, and countercultural edge becomes inevitable. This, in turn, renders *making* into a new form of hierarchy legitimized by a new breed of well-educated individuals operating either as replacements for, or supplements to the "dominant ideational regime." This is particularly the case with Maker entrepreneurs proper.

Maker entrepreneurship proper in Shenzhen, then, has become an experiment and instrument not only for building a stable national economy, but also for becoming a key player in the global economy—insofar as possible on China's terms. It has contributed to the formation of a Chinese "dominant ideational regime" that has been instrumental in rendering Shenzhen a national and global model city positioned as a distinctive, innovative IoT powerhouse, something that is reflected in China's new status as the top filer of patents worldwide, and its development of economic and political alliances both nationally and internationally. Where non-entrepreneurial Maker communities in Shenzhen embody protest through undertaking small-scale, long-term attempts to improve life conditions for ordinary citizens, we argue that the said "regime" appears to be protesting Western dominance by pursuing not a fundamentally different world order—but one tangled up in colonialist, imperialist, market-socialist, and freewheeling, neoliberal approaches. These approaches engender new forms of hierarchies and exacerbate old conditions of existence that Maker entrepreneurship proper is increasingly implicated in. This points to "old," "new," and "uncertain" futures on the horizon.

NOTES

1. Neil Gershenfeld, *Fab: The Coming Revolution on Your Desktop—From Personal Computers to Personal Fabrication* (New York: Basic Books, 2008).
2. Mark Hatch, *The Maker Movement Manifesto: Rules for Innovation in the New World of Crafters, Hackers, and Tinkerers* (New York: McGraw Hill Education, 2013).
3. Recent discourses speak of a "fourth industrial revolution" to capture the latest technological advancements propelling the digital revolution through the development of data platforms deployed across various industries. See also Jia and Nieborg as well as Lin and de Kloet in this volume. Worth looking at also are Klaus Schwab, *The Fourth Industrial Revolution* (London: Penguin Books, 2017); Naubahar Sharif and Yu Huang, "Introduction: Innovation and Work in East Asia," *Science, Technology & Society* 24, no. 2 (2019): 193–98.
4. Adrian Smith, Sabine Hielscher, Sascha Dickel, Johan Sorderberg, and Ellen van Oost, *Grassroots Digital Fabrication and Makerspaces: Reconfiguring, Relocating and Recalibrating Innovation*, Science and Technology Policy Research, Working Paper Series (Brighton: University of Sussex, 2013), 4.
5. Caitlin Bagley, *Makerspaces: Top Trailblazing Projects* (Chicago: American Library Association, 2014); Kimberly Sheridan, Erica R. Halverson, Breanne K. Litts, Lynette Jacobs-Priebe, Trevor Owens, "Learning in the Making: A Comparative Case Study of Three Makerspaces," *Harvard Educational Review* 84, no. 4 (2014): 505–31.
6. Tom Saunders and Jeremy Kingsley, *Made in China: Makerspaces and the Search for Mass Innovation* (Nesta and British Council, 2016), 20, https://www.nesta.org.uk.
7. Adrian Smith, Mariano Fressoli, Dinesh Abrol, Elisa Arond, and Adrian Ely, *Pathways to Sustainability: Grassroots Innovation Movements* (London: Routledge, 2016); Chris Anderson, "Makers and DIY Manufacturing," *ParisTech Review* (January 4, 2013).
8. Chris Anderson, *Makers: The New Industrial Revolution* (London: Random House Business, 2012).
9. James Fallows, "Why the Maker Movement Matters: Part 1–The Tools Revolution," *The

Atlantic, June 5, 2016; Bruce E. Massis, "What's New in Libraries: 3D Printing and the Library," *New Library World* 114, no. 7/8 (2013): 351–54.

10. Dane Stangler and Kate Maxwell, "DIY Producer Society," *Innovations: Technology, Governance, Globalization* 7, no. 3 (2012): 3–10.

11. Catarina Mota, "The Rise of Personal Fabrication," *Proceedings of the 8th ACM Conference on Creativity and Cognition* (New York: ACM, 2011): 279–88.

12. Smith, Hielscher, Dickel, Sorderberg, and van Oost, *Grassroots Digital Fabrication and Makerspaces*; Heather M. Moorefield-Lang, "Makers in the Library: Case Studies of 3D Printers and Makerspaces in Library Settings," *Library Hi Tech* 32, no. 4 (2014): 583–93.

13. For instance, see Jeremi Suri, "The Rise and Fall of an International Counterculture, 1960–1975," *American Historical Review* 114, no. 1 (February 2009): 45–68.

14. Shaowen Bardzell, Jeffrey Bardzell, and Sarah Ng, "Supporting Cultures of Making: Technology, Policy, Visions, and Myths," *CHI 2017: Proceedings* (ACM Press, 2017), 6523–35, http://dx.doi.org/10.1145/3025453.3025975; Sarah R. Davies, *Hackerspaces: Making the Maker Movement* (Malden, MA: Polity Press, 2017).

15. One perception, for instance, views the international countercultural struggles as aberrations that led to major disruptions of orderly life and social norms. We disagree and argue instead that these struggles—including the "Great Proletarian Cultural Revolution in China" (1966–1976)—were connected by a set of aspirations and values that advocated a reimagining of conditions of existence through transforming established hierarchies and formalities dominating societal relations. In their different ways, the struggles strove for political mobilization, experimentation, egalitarianism, less state direction and more participatory governance, pragmatism, and economic prosperity for all, among other things. Although only some changes materialized—albeit to varying degrees—they came at a huge cost in terms of human suffering, death, and massive material and immaterial displacement in many parts of the "Third World" including China. See Arthur Marwick, *The Sixties: Cultural Revolution in Britain, France, Italy, and the United States, 1958–1974* (New York: Oxford University Press, 1998); Xing Li, "The Chinese Cultural Revolution Revisited," *China Review* 1, no. 1 (Fall 2001): 137–65; Kristin Ross, *May '68 and Its Afterlives* (Chicago: University of Chicago Press, 2002); Steve Giles and Maike Oergel, *Counter-cultures in Germany and Central Europe: From Sturm und Drang to Baader-Meinhof* (Oxford: Peter Lang, 2003); Shao Yinong and Mu Chen, "It Is Not Merely a Memory," in *Burden or Legacy: From the Chinese Cultural Revolution to Contemporary Art*, ed. Jiehong Jiang (Hong Kong: Hong Kong University Press, 2007), 85–91; Tamara Chaplin and Jadwiga E. Pieper Mooney, eds., *The Global 1960s: Convention, Contest, and Counterculture* (Abingdon, UK:

16. John Case and Rosemary C. R. Taylor, Co-ops, Communes, and Collectives: Experiments in Social Change in the 1960s and 1970s (New York: Pantheon, 1979).
17. Versluis, "On Counterculture," 147, 151.
18. Bardzell, Bardzell, and Ng, "Supporting Cultures of Making," 6531; Davies, *Hackerspaces*, 160.
19. To differing degrees, countercultural activity and *making* display a devotion to social justice, understood in this context as the urge to help improve society by leveraging resources and tools to transform lives for the better.
20. Roszak, *The Making of a Counterculture*; Andrew Kirk, "Appropriating Technology: The Whole Earth Catalog and Counterculture Environmental Politics," *Environmental History* 6, no. 3 (2001): 374–94; Fred Turner, *From Counterculture to Cyberculture: Stewart Brand, the Whole Earth Network, and the Rise of Digital Utopianism* (Chicago: University of Chicago Press, 2006); Hatch, *The Maker Movement Manifesto*; Adrian Smith and Andrew Stirling, "Innovation, Sustainability and Democracy: An Analysis of Grassroots Contributions," *Journal of Self-Governance and Management Economics* 6, no. 1 (2018): 64–97; David Swift, *A Left for Itself: Left-wing Hobbyists and Performative Radicalism* (Hampshire, UK: Zero Books, 2019).
21. Thomas Frank, *The Conquest of Cool: Business Culture, Counterculture, and the Rise of the Hip* (Chicago: University of Chicago Press, 1997); Matt Ratto, "Critical Making: Conceptual and Material Studies in Technology and Social Life," *Information Society* 27, no. 4 (2011): 252–60; Yochai Benkler and Helen Nissenbaum, "Commons-Based Peer Production and Virtue," *Journal of Political Philosophy* 14, no. 4 (2006): 394–419; Chaplin and Pieper Mooney, "Introduction," 1–11.
22. Dieter Rucht, "Protest Cultures in Social Movements," in *Protest Cultures: A Companion*, ed. Kathrin Fahlenbrach, Martin Klimke, and Joachim Scharloth (New York: Berghahn Books, 2016), 28.
23. See Colin Marshall, "Learning from Las Vegas: What the Strip Can Teach Us about Urban Planning," *The Guardian*, February 9, 2015; British Council, "Say Hello to China's Maker Capital," http://creativeconomy.britishcouncil.org/blog/16/10/12/say-hello-chinas-maker-capital/; Jamie Carter, "A Geek's Guide to Shenzhen, the Global Gadget Capital," *TechRadar*, May 2, 2016, http://www.techradar.com/news/world-of-tech/a-geek-s-guide-to-shenzhen-the-global-gadget-capital-1320107; *Shenzhen: The Silicon Valley of Hardware*, https://www.youtube.com/watch?v=SGJ5cZnoodY; Ezra F. Vogel, "Foreword," in *Learning from Shenzhen: China's Post-Mao Experiment from Special Zone to Model City*, ed. Mary Ann O'Donnell, Winnie Wong, and Jonathan Bach

(Chicago: University of Chicago Press, 2017), vii; Mary Ann O'Donnell, Winnie Wong, and Jonathan Bach, eds., "Introduction: Experiments, Exceptions, and Extensions," in *Learning from Shenzhen*, 2, 12.

24. Fallows, "Why the Maker Movement Matters: Part 1–The Tools Revolution"; Silvia Lindtner and David Li, "Created in China: The Makings of China's Hackerspace Community," *Community + Culture Forum* (2012): 18–22.

25. Anthony Townsend, Lyn Jeffery, Devin Fidler, and Mathias Crawford, "The Future of Open Fabrication," *Technology Horizons Program* (Institute for the Future, 2011); Saunders and Kingsley, *Made in China*; David Gauntlett, *Making Is Connecting: The Social Power of Creativity, from Craft and Knitting to Digital Everything*, 2nd ed. (Cambridge: Polity Press, 2018).

26. See Lindtner and Li, "Created in China"; and Luisa Mengoni, "The Pirates and the Makers/Shanzhai," Victoria and Albert Museum (2016), https://www.vam.ac.uk/blog/ for a discussion of the semantic development of the term. It is now widely understood to mean a large bottom-up ecosystem underpinned by collaborative innovation that is rapid, flexible, and open mostly without regard for IP issues. See also Miao Lu in this volume.

27. David Li, "The New Shanzhai: Democratising Innovation in China," *ParisTech Review* (2014), http://www.paristechreview.com/2014/12/24/shanzhai-innovation-china/?media=print/.

28. Luisa Mengoni, "From Shenzhen: Shanzhai and the Maker Movement," Victoria and Albert Museum (2015), https://www.vam.ac.uk/blog/; Clive Thompson, "How a Nation of Tech Copycats Transformed into a Hub for Innovation," *Wired* (2015), https://www.wired.com/2015/12/tech-innovation-in-china/.

29. Mengoni, "The Pirates and the Makers/Shanzhai." Also in this volume, Miao Lu provides an interesting account of this phenomenon. Lu discusses how a Shenzhen-based company called Transsion Holdings Ltd leverages innovation to cater to the needs and wants of so-called "bottom of the pyramid" (BOP) markets in Ghana and Nigeria among other countries in Africa.

30. Saunders and Kingsley, *Made in China*, 8.

31. Thompson, "How a Nation of Tech Copycats Transformed into a Hub for Innovation."

32. Silvia Lindtner, Anna Greenspan, and David Li, "Shanzhai: China's Collaborative Electronics-Design Ecosystem," *The Atlantic*, May 18, 2014.

33. Saunders and Kingsley, *Made in China*, 8.

34. Naubahar Sharif and Yu Huang, "Achieving Industrial Upgrading through Automation in Dongguan, China," *Science, Technology & Society* 24, no. 2 (2019): 237–53. See also

35. Hoyng and Chong in the introduction of this volume.
35. Boy Lüthje, "Platform Capitalism 'Made in China'? Intelligent Manufacturing, Taobao Villages and the Restructuring of Work," *Science, Technology & Society* 24 no. 2 (2019): 199–217. Again, we point to the work of Jia and Nieborg, as well as Lin and de Kloet in this volume.
36. Saunders and Kingsley, *Made in China*.
37. Silvia Lindtner, "Hackerspaces and the Internet of Things in China: How Makers are Reinventing Industrial Production, Innovation, and the Self," *China Information* 28, no. 2 (2014): 145–67.
38. Li, "The New Shanzhai"; Luisa Mengoni, "From Maker Spaces to Making Culture," Victoria and Albert Museum (2015), https://www.vam.ac.uk/blog/international-initiatives/from-maker-spaces-to-making-culture.
39. Marianna Cerini, "DIY Nation: China's New Wave of Young Innovators," *That's* magazine, February 5, 2015, http://www.thatsmags.com/china/post/8675/diy-nation-the-fast-growing-world-of-chinas-hackerspace-communities.
40. Saunders and Kingsley, *Made in China*.
41. Thompson, "How a Nation of Tech Copycats Transformed into a Hub for Innovation."
42. Smith, Hielscher, Dickel, Sorderberg, and van Oost, *Grassroots Digital Fabrication and Makerspaces*, 10.
43. Robert E. Stake, "Case Studies," in *Handbook of Qualitative Research*, ed. Norman K. Denzin and Yvonna S. Lincoln, 2nd ed. (Thousand Oaks, CA: Sage, 2000), 435–54.
44. See, for example, Catherine Durose, Yasminah Beebeejaun, James Rees, Jo Richardson, and Liz Richardson, *Towards Co-Production in Research with Communities* (Connected Communities) (Arts and Humanities Research Council, 2012); Sarah Banks, Angie Hart, Kate Pahl, and Paul Ward, *Co-producing Research: A Community Development Approach* (Bristol, UK: Policy Press, 2019).
45. Martyn Hammersley and Paul Atkinson, *Ethnography: Principles in Practice*, 3rd ed. (London: Routledge, 2007).
46. See, for instance, : https://makerbay.net/zh/shenzhen-makerspaces-update-nov-2017; and "6 Coolest Makerspaces in Shenzhen, China," https://getinthering.co/6-coolest-makerspaces-of-shenzhen-china/
47. People's Republic of China, "Mass Entrepreneurship and Innovation as New Growth Engine," March 3, 2016, http://english.www.gov.cn/premier/news/2016/03/03/content_281475300571752.htm.
48. Pamela Ambler, "This Chinese City Is the Mecca of Hardware Startups," *Forbes India* (2018), https://www.forbesindia.com/article/cross-border/

this-chinese-city-is-the-mecca-of-hardware-startups/49897/1.

49. Ambler, "This Chinese City is the Mecca of Hardware Startups"; Andrea Rossi, "Why Shenzhen is the Mecca of Hardware," *Road to China* (2019), http://www.road-to-china.com/why-shenzhen-is-the-mecca-of-hardware/.

50. See, for instance, "Shenzen Map for Makers," http://collections.vam.ac.uk/item/O1282473/shenzhen-map-for-makers-map-seeed-studio/; Joshua Bateman, "Life inside a Shenzhen Hardware Accelerator," https://www.fastcompany.com/3066987/life-inside-a-shenzhen-hardware-accelerator; O'Donnell, Wong, and Bach, eds., *Learning from Shenzhen*.

51. We learned that independent design houses and industrial design companies tend to maintain an up-to-date record of existing design templates of all kinds of products on the market. In many instances, designs of very popular products tend to be the main draw because of their promise of profit, as opposed to new, untested designs that tend to be considered as risky.

52. This model describes a learning environment in which learners acquire much of their knowledge and experience through proactive exploration of issues and questions around them, interacting with peers about these, and obtaining some support from formal learning. It is most commonly known as "project-based learning" and applied as an approach that foregrounds experiential learning. It is an integral component of Maker Ed.

53. To varying degrees, the case study sites are furnished with the standard equipment and infrastructure commonly associated with makerspaces and Fab Labs. Examples of equipment and resources that we observed at each of the sites included 3D printing, 3D scanning, Computer Numerical Control (CNC) machining, laser cutting, vinyl cutting, precision milling, carpentry machinery, electronics, prototyping tools, workstations, and a variety of workshop kit among many others. CNC machining is understood as a system used in "digital" or "smart" manufacturing whereby computers operate the tools and machines. Precision milling is a technique that involves working with a digitally coded system that instructs milling machines to move, cut, and drill materials at different angles as accurately as possible. Skills training is offered to enable users to operate equipment appropriately. For illustration, Figure 1 in this chapter shows equipment in the 3D Printing Room at Chaihuo x.factory (previously known as Chaihuo Maker Space).

54. See Chong-Moon Lee, William F. Miller, Marguerite Gong Hancock, and Henry S. Rowen, eds., *The Silicon Valley Edge: A Habitat for Innovation and Entrepreneurship* (Stanford, CA: Stanford University Press, 2001); Boy Lüthje, Stefanie Hürtgen, Peter

Pawlicki, and Martina Sproll, *From Silicon Valley to Shenzhen: Global Production and Work in the IT Industry* (Lanham, MD: Rowman & Littlefield, 2013); Cara Wallis and Jack Linchuan Qiu, "Shanzhaiji and the Transformation of the Local Mediascape in Shenzhen," in *Mapping Media in China: Region, Province, Locality*, ed. Wanning Sun and Jenny Chio (Abingdon, UK: Routledge, 2012), 115; Yutao Sun and Seamus Grimes, *China and Global Value Chains: Globalization and the Information and Communications Technology Sector* (London: Routledge, 2018), 115.

55. Kat Braybrooke and Adrian Smith, "Editors' Introduction: Liberatory Technologies for Whom? Exploring a New Generation of Makerspaces Defined by Institutional Encounters," *Journal of Peer Production*, no. 12 (2018): 4.

56. Bardzell, Bardzell, and Ng, "Supporting Cultures of Making"; Jeremy Hunsinger and Andrew Schrock, "The Democratization of Hacking and Making," *New Media & Society* 18, no. 4 (2016): 535–38.

57. Kunal Sinha, China's Creative Imperative: How Creativity Is Transforming Society and Business in China (Singapore: Wiley, 2008); John Naisbitt and Doris Naisbitt, China's Megatrends: The Eight Pillars of a New Society (New York: HarperCollins, 2010); Edmund Phelps, Mass Flourishing: How Grassroots Innovation Created Jobs, Challenge, and Change (Princeton, NJ: Princeton University Press, 2013).

58. Lüthje, Hürtgen, Pawlicki, and Sproll, *From Silicon Valley to Shenzhen*; Neil Gershenfeld, Alan Gershenfeld, and Joel Cutcher-Gerschenfeld, *Designing Reality: How to Survive and Thrive in the Third Digital Revolution* (New York: Basic Books, 2017); Pip Shea and Xin Gu, "Makerspaces and Urban Ideology: The Institutional Shaping of Fab Labs in China and Northern Ireland," *Journal of Peer Production*, no. 12 (2018): 78–91.

59. Davies, *Hackerspaces*, 172.

60. Lorraine Justice, *China's Design Revolution* (Cambridge, MA: MIT Press, 2012); Andrew Chubb, "China's Shanzhai Culture: 'Grabism' and the Politics of Hybridity," *Journal of Contemporary China* 24, no. 92 (2015): 260–79; John Hartley, Wen, and Henry Siling Lin, *Creative Economy and Culture: Challenges, Changes and Futures for the Creative Industries* (London: Sage, 2015).

61. Cindy Kohtala, "The Socialmateriality of Fab Labs: Configurations of a Printing Service or Counter-Context?" *Journal of Peer Production*, no. 12 (2018): 92–110; Xin Gu, "The Paradox of Maker Movement in China," in *Making Our World: The Hacker and Maker Movements in Context*, ed. Jeremy Hunsinger and Andrew Schrock (New York: Peter Lang, 2019), 271–91; Chen-Yi Lin, "The Hybrid Gathering of Maker Communities in Taipei Makerspaces: An Alternative Worlding Practice," *City, Culture and Society* 19 (2019): 1–8.

62. Gauntlett, *Making Is Connecting*, 1. As reflected by Figure 2 in this chapter, the motto of the Second Shenzhen International Maker Week in 2017 reads "Small Creativity, Big World," which speaks to "everyday inventiveness" as one of the central themes of this volume.

63. See also Wen, "Making in China: Is Maker Culture Changing China's Creative Landscape?" *International Journal of Cultural Studies* 20, no. 4 (2017): 343–60.

64. Lyn Jeffrey, "Innovation Spaces of the Future: Research Notes on China's Shanzhai Meeting the Makers" (2011), http://www.iftf.org/node/3943; Silvia Lindtner, "Hacking with Chinese Characteristics: The Promises of the Maker Movement against China's Manufacturing Culture," *Science, Technology & Human Values* 40, no. 5 (2015): 854–79.

65. Lindtner, Greenspan, and Li, "Shanzhai"; Saunders and Kingsley, *Made in China*; Xianyue Li, "Another Eye to Inspect the Cultural Revolution in China," *Perspectives on Global Development and Technology* 16 (2017): 260–73; Gu, "The Paradox of Maker Movement in China."

66. Jennifer Tyree-Hageman, "From Silicon Valley to Wall Street: Following the Rise of an Entrepreneurial Ethos," *Berkeley Journal of Sociology* 57 (2013): 74–113; Gershenfeld, Gershenfeld, and Cutcher-Gerschenfeld, *Designing Reality*; O'Donnell, Wong, and Bach, eds., *Learning from Shenzhen*; Wen, "Making in China"; Hallam Stevens, "The Quotidian Labor of High Tech: Innovation and Ordinary Work in Shenzhen," *Science, Technology & Society* 24, no. 2 (2019): 218–36.

67. Neil Gershenfeld, "How to Make Almost Anything: The Digital Fabrication Revolution," *Foreign Affairs* 91, no. 6 (2012): 43–57; Jeremy Hunsinger and Andrew Schrock, eds., "Introduction," in *Making our World: The Hacker and Maker Movements in Context* (New York: Peter Lang, 2019), vii–xii.

68. Seeed, https://www.seeedstudio.com/Crazyflie-2.0---p-2103.html.

69. "Tiny, Hackable Quadcopter," Wired.com, https://www.wired.com/2013/02/crazyflie-nano/.

70. See "Make Your Ideas Fly," Bitcraze, https://www.bitcraze.io/.

71. Naisbitt and Naisbitt, *China's Megatrends*; Rossi, Marshall, and Julier, *China's Creative Communities*; Gu, "The Paradox of Maker Movement in China."

72. Tyree-Hageman, "From Silicon Valley to Wall Street," 104.

73. Case and Taylor, *Co-ops, Communes, and Collectives*; AnnaLee Saxenian, *Regional Advantage: Culture and Competition in Silicon Valley and Route 128* (Cambridge, MA: Harvard University Press, 1994); Marc A. Smith and Peter Kollock, eds., *Communities in Cyberspace* (London: Routledge, 1999); Timothy J. Sturgeon, "How Silicon Valley Came to Be," in *Understanding Silicon Valley: The Anatomy of an Entrepreneurial Region*, ed.

Martin Kenney (Stanford, CA: Stanford University Press, 2000), 15–47; Lee, Miller, Hancock, and Rowen, eds., *The Silicon Valley Edge*; Turner, *From Counterculture to Cyberculture*; Anderson, *Makers*; Tyree-Hageman, "From Silicon Valley to Wall Street."

74. Wallis and Qiu, "Shanzhaiji and the Transformation of the Local Mediascape in Shenzhen," 117.
75. See Case and Taylor, *Co-ops, Communes, and Collectives*; Justice, *China's Design Revolution*; O'Donnell, Wong, and Bach, eds., *Learning from Shenzhen*; The White House, "Nation of Makers," https://obamawhitehouse.archives.gov/nation-of-makers.
76. Michael Keane, *Creative Industries in China: Art, Design and Media* (Hoboken, NJ: Wiley, 2013); Hartley, Wen, and Lin, *Creative Economy and Culture*; Shea and Gu, "Makerspaces and Urban Ideology"; Gu, "The Paradox of Maker Movement in China."
77. See, for instance, Lee, Miller, Hancock, and Rowen, eds., *The Silicon Valley Edge*; Divya Leducq, "Self-Made Man," in *Encyclopedia of Creativity, Invention, Innovation and Entrepreneurship*, ed. Elias G. Carayannis (New York: Springer, 2013), 1604–7.
78. WIPO, "China Becomes Top Filer of International Patents in 2019," https://www.wipo.int/pressroom/en/articles/2020/article_0005.html.
79. Shea and Gu, "Makerspaces and Urban Ideology," 84, 85.
80. Braybrooke and Smith, "Editors' Introduction: Liberatory Technologies for Whom?," 2.
81. Chaplin and Pieper Mooney, "Introduction," 2.

Platformization of the Unlikely Creative Class

Kuaishou and Chinese Digital Cultural Production

Jian Lin and Jeroen de Kloet

Lonely, I feel alive, I just wanna touch the sky. And you, girl please don't cry. ... Together we sing a song that will take me to your heart!"¹ These words are uttered by a young man in a black sleeveless T-shirt sitting in front of a computer screen; Tian You is his name, and he calls himself an MC.² In a recorded live-streaming video, he expresses his anger at the prevalence of discrimination, the unequal distribution of wealth, and social inequality in the form of *Hanmai* (喊麦),³ a Chinese rap-like performance that has been popular on the internet since 2014. Thanks to live-streaming platforms like YY and *Kuaishou* (快手), its particular combination of coarse narration and rhythmical music is now celebrated by millions of young Chinese. Not long ago, Li Tianyou, which is Tian You's real name, was a scrawny high-school dropout struggling to make a living in a small, dreary industrial city in northeastern China. Since 2014, he has been one of the best-known Chinese "Internet celebrities,"⁴ commanding a fan base of over 35 million people for his live-streaming shows on Kuaishou and earning more than €1.8 million a year in payments from his fans and advertisers. And Tian You is not alone. Enabled by emerging Chinese digital platforms, thousands of young Chinese like him are posting images and short videos, and making live-streaming shows to flaunt their creative talents while also hoping to earn a lot of money. Most of them are uneducated young Chinese from small cities and rural areas. They earn an average monthly income ranging from RMB2,000 (€250) to RMB4,000 (€500); successful ones can earn as much as 1 million (€120,000) per month.[5]

Just as the editors of this book point out in the introduction, however, technical innovation and its financialization bring along both opportunities and great uncertainties and risks. The new form of creative business facilitated by Kuaishou also comes with risks, though for a different reason. The ranting style of performance and its enormous popularity with massive online fan bases have also troubled the Chinese authorities, just as rap music and black pop culture have done in American society.[6] In early 2018, Tian You was accused by China Central Television, the central television network controlled by the state, of talking about pornography and drugs during his live-streaming. Shortly after, Tian You and some other top-ranked live-streamers were banned by all Chinese platforms, and their performing careers seemed to have come to an end.[7]

When thinking about the "creative class," one tends to imagine an urban elite, an educated group of predominantly young people who work in the cultural industries and gather in hipster bars, dressed in the latest local and cosmopolitan designer brands. But, as the story of Tian You shows, the emerging digital and platform economy also offers opportunities for lower educated, more marginal people to participate as producers in the Chinese creative economies. According to the White Paper on Chinese Digital Economy 2016 released by the Cyberspace Administration of China (CAC), the national administrative bureau in charge of the Chinese internet communication sector, China's digital economic aggregate in 2016 reached RMB226 billion, constituting 30.3 percent of China's Gross Domestic Product (GDP).[8] More importantly, the convergence of traditional sectors and digital internet technology has replaced the ICT (information, communication, and technology) manufacturing, telecommunication, and software industries to become the "main engine" of the Chinese digital economy.[9] Various digital platforms such as Taobao and Wechat have played a crucial role in such convergence processes, forming the so-called platform economy.[10] In the media and cultural sectors, digital convergence has contributed 45.4 percent of the total economic growth in the broadcast, television, film, and recording industries. According to Nieborg and Poell, such platformization marks "the penetration of economic, governmental, and infrastructural extensions of digital platforms into the web and app ecosystems, fundamentally affecting the operations of the cultural industries."[11] Data-based digital/internet technologies afford platforms like Kuaishou a high degree of connectivity that allows them to mediate between various actors, including content producers, end-users, and advertisers, and to incorporate them into the platform-dominated network system of "the multi-sided markets."[12] The platformization of cultural production blurs the boundaries between traditional media forms and gives rise to an exponential growth of user-generated content production. The multisided network system not only enables traditional media companies to expand their content business, but also, as Tian You's story shows, produces opportunities for marginalized individuals to become self-employed "creative workers."

In this chapter, we want to investigate this emerging yet "unlikely" creative class in China, which is part of the rapid platformization of Chinese cultural production, and engage with the aesthetics of the work this class produces. How are these diverse and sometimes marginal groups of individuals and their creativities mobilized and incorporated into the platform creative economy? What kinds of aesthetics and culture are produced on these content platforms? How does platformization relate to the Chinese state's governance of culture, economy, and society? And what are the differences and similarities between Chinese platformed cultural production and its counterpart in "the West"? To address these questions, the chapter focuses on one particular platform, Kuaishou. Labeled by Chinese mainstream media as "revitalising Chinese rural culture,"[13] the app attracts hundreds of millions of Chinese from the countryside and the second- and third-tier cities. Since its launch in 2012, it has become one of the most popular video-sharing platforms in China, allowing its users not only to watch, make, and distribute various genres of short videos, but also to become "complementors" of the platform: professional content producers contributing to the platformization of cultural production in China.[14] Before moving to our analysis, we first discuss the Chinese processes of platformization.

This chapter starts with an introduction to the issue of digital labor, the political economy of the Chinese platform creative economy (see a more comprehensive discussion of this issue in the chapter by Jia and Nieborg), and the specific position of Kuaishou in this system. We distance ourselves from viewing digital labor solely in terms of exploitation and precarity. We consider Kuaishou a strong example of multipolar platformization; it presents a case different from platforms like Instagram. As we will show, the platform allows for a digital entrepreneurship among an unlikely class that includes migrant workers and farmers. We thus steer away in this chapter from the focus on risk and uncertainty, as analyzed in the introduction under the disruption/structure paradox. While the technologies and the political economy are integrated both nationally and globally, the actual uses also cause moments of differentiation.

We will show how the Chinese platform cultural economy distinguishes itself from its Western counterparts in its special state-platform relations, which simultaneously promote and limit platformization. The close link with the state is seen to constitute a third dimension of contingency, in addition to "platform dependence" and "contingent commodities," identified by Nieborg and Poell as the forms of contingency characterizing the Western platform cultural economy, which we will elaborate on below. The following section analyzes the workings of the Kuaishou platform.[15] Using the "walkthrough" method of Light, Burgess, and Duguay,[16] "a way of engaging directly with an app's interface to examine its technological mechanisms and embedded cultural references to understand how it guides users and shapes their experiences," we examine how the contingent platform business induced by the complicated state-commerce relationship is encoded in the algorithms of Kuaishou. Finally, to probe the characteristics of this unlikely creative class and the specific aesthetics they produce, we analyze 200 trending videos and the everyday user activities of 20 popular Kuaishconou accounts.[17] Besides conducting a visual and digital analysis of the videos, we held fourteen in-depth interviews with managers from the Kuaishou company, content producers, algorithm engineers, and other professionals whose work is related to Kuaishou and the Chinese platform creative economy. We argue that the platformization of cultural production in China accommodates the state's "entrepreneurial solutionism," while also producing a digital creative entrepreneurship among Chinese "grassroots individuals" and a dynamic digital culture permeated with contingency and negotiation.

Thus, Kuaishou, as a case of multipolar platformization, guides our attention away from a univocal focus on risk in the disruption versus structure paradox, just as it is a case of both global (and national) integration, as well as of differentiation. The latter constitutes the core of this chapter; we show how Kuaishou allows for the emergence of an unlikely creative class. While we zoom in on Kuaishou as a Chinese platform, we believe our findings speak back to both creative labor and platform studies in general. Studies on creative labor in, say, the United States predominantly focus on an urban creative class. In the wake of Richard Florida's work, cities around the globe aspire to become creative.[18] In Richard Lloyd's work on the creative class in Chicago, he shows how a neo-bohemian ideology helps obscure the precariousness of creative work.[19] While people believe themselves to be free and creative, they are often abused and unfree. What our study adds to this debate are second- and third-tier cities as well as rural areas. Kuaishou thus helps to debunk the singular focus on the urban in creative labor studies. In the domain of platform studies, our analysis shows how both a different political economy and a different demographic of a platform's target audience shape a quite different mode of platformization. We witness an everyday inventiveness that propels a technodiversity that asks for a recalibration of platform studies.

Digital Labor and the Chinese Platformed Cultural Economy

As a global phenomenon, the platform economy has been extensively criticized for the type of labor it involves. Van Doorn, for example, notes that in the platform economy, contracted labor has been replaced by "platform labor," which adopts "a more austere and zero-liability peer-to-peer model that leverages software to optimize labor's flexibility, scalability, tractability, and its fragmentation."[20] In this sense, workers are regarded as complementors or subcontractors, instead of employees, of the platform companies, which are therefore exempted from providing labor protection. Critical political economists have also attacked content-based platforms for deliberately inviting users to become "prosumers" and thus contributing to the exploitation of free, creative labor.[21]

Although these arguments provide valuable insights into the new labor conditions in the global platform economy, they tend to overlook the active agency or personal practices of digital/platform creators. The "multi-sided markets" of platform businesses suggest a more complicated relationship among different actors in the operation of platformization. The networked mode of cultural production indicates that "the costs of the production and consumption of goods and services" will affect other complementors of the platform such as content producers and advertisers, and vice versa.[22] As the word "complementor" implies, the commercial relationship between platform companies and complementors is not only exploitative, but also collaborative and symbiotic. The long-term financial success of digital platforms is thus not simply based on the exploitation of platform labor, but is contingent upon commercial collaboration between platform companies, content producers, and other complementors. In the case of Kuaishou, as we will show in the following sections, by actively utilizing the digital system afforded by the platform, "grassroots" content producers are enabled to develop a digital creative entrepreneurship.

In the development of the platform economy, the Chinese state is a crucial agent. We find resonances here with other localities, such as Turkey, in which President Erdogan also promotes creativity and innovation, as discussed in the chapter by Sezgin and Binark in this volume. And while Mutibwa and Xia in this volume show how makerspaces in Shenzhen allow for a countercultural logic, such a logic is much less likely when it concerns a platform like Kuaishou. Scale is a key marker of difference here—small pockets of resistance are more likely to survive when compared to nationwide platforms. As Yu Hong illustrates, the Chinese government has pledged to place information and communication at the center of the national economic restructuring plan, using ICT as industries and infrastructures to transform traditional industrial sectors.[23] However, as Tian You's experience shows, the state wants not only to "profit" from information and culture, but also to control and shape it so as to maintain social and political stability. With regard to Kuaishou, therefore, we need to begin by asking how this platform's cultural economy is governed by the Chinese state. How does state governance affect the working experience of the various platform content creators?

In 2015, Prime Minister Li Keqiang announced China's "Internet+" agenda.[24] This is a new national development strategy that aims at boosting and restructuring the national economy through the upgrading of digital infrastructure and technological innovation.[25] "Internet+" is the continuation of the state's economic restructuring plan, which aims to replace the unsustainable "export-driven," "investment-dependent" model with a "consumption-based" and "innovation-driven" economy. The new policy agenda puts the internet at the center, aiming to integrate network connectivity and the "disruptive business and managerial model" (of decentralized, private, post-Fordist corporate management) with a wide range of traditional sectors, from manufacturing, agriculture, energy, finance, and transportation to public services and education.[26] Moreover, the "Internet+" strategy pledges to propel a new digital economy that can foster and benefit small start-ups, entrepreneurship, and innovation. As such, it dovetails with another policy agenda championed by the state government under the name "Mass Entrepreneurship and Innovation" (大众创业万众创新 *dazhong chuangxin, wanzhong chuangye*), which has contributed to the "Maker Movement" in Shenzhen noted in Mutibwa and Xia's chapter.[27] The latter policy seeks to mobilize the creativity and innovative power of grassroots individuals for national economic growth. "Internet+" complements the "Mass

Entrepreneurship" strategy in the sense that the prosperous digital economy provides opportunities for grassroots individuals to find employment and become entrepreneurs. According to Premier Li Keqiang,

> Internet+ not only produces new economic driving power, but also creates the largest platform for the sharing economy, which stages "Mass Entrepreneurship and Innovation" and will deeply affect our economy, society and everyday life. It provides opportunities for not only techno elites and entrepreneurs, but also millions of *caogen* (草根 grassroots individuals) to exploit their talent and to realise their special value.[28]

In practice, as the official statistics cited earlier indicate, the state agenda of "Internet+" and "Mass Entrepreneurship" has greatly contributed to the surging platformed creative economy in China. Kuaishou, together with its competitors such as Toutiao and Douyin,[29] enables both traditional media companies and Chinese "grassroots individuals" to establish and expand their content business.[30] Echoing the analysis in the introduction of the book, Kuaishou's platform and business ecology is characterized by a model of technological innovation that is driven by both state intervention and financialization.

Launched in 2012, Kuaishou is an algorithm-based video and live-streaming platform that allows registered users to create and post all kinds of short videos online. These videos show activities ranging from cooking, body building, skills training, and applying makeup to micro fiction films. The remarkably diverse content made by millions of online users is computed and pushed to targeted viewers by Kuaishou's algorithm recommendation system. This algorithm system, as Gillespie suggests,[31] replaces the role of traditional editors in the selection and distribution of content, providing a seemingly more "objective" model based on the artificial-intellectual (AI) computation of user data rather than on editors' "subjective" preferences.[32] The most important distinguishing characteristic of Kuaishou is that the majority of its users consist of rural or third- and fourth-tiered city-based, uneducated young Chinese.[33] As we will show, Kuaishou enabled this group to become an "unlikely creative class," actively performing their vernacular creativity through self-taught skills.[34] In addition, they use the digital system of Kuaishou to monetize their creative production through advertising and e-commerce. At first sight, Kuaishou's platform content business and its "unlikely creative class" seem to fit comfortably with the state's expectation of "mass entrepreneurship."

However, the challenge for Kuaishou is that its user-generated content has to be in line with the authorities' expectations of "what kind of stories should be told." This is especially challenging because the stakes are high: "Internet+" is not just about "restructuring the economy," but also about restructuring culture and society. The Chinese authorities have been eager to promote a carefully curated national imagery to wield "soft power" on the global stage on the one hand, while expecting a conforming culture that ensures social stability and national unity on the other. As Wanning Sun highlights, this refers to a double agenda: to "globally present [China] as a player whose values, ethics, and sensibilities are compatible with . . . its international counterparts," while domestically "avoid[ing] 'chaos' at all cost, including heavy-handed censorship, in order to ensure social stability and national unity."[35] This double agenda applies to Chinese digital platforms. The platformed cultural production system puts users at the center of production, endowing content producers with more autonomy. Yet, as long as these platforms operate domestically, they are not immune to censorship or the state's demand for a compliant culture. According to the CAC, all types of content providers should "abide by the law, adhere to the correct values, and help disseminate socialist core values and cultivate a positive and healthy on-line culture."[36] As the central supervisory entity for the Chinese internet communication sectors, the CAC is a powerful government agency under the leadership of the Central Cyberspace Affairs Commission, headed directly by the Chinese president Xi Jinping. Founded in 2014, the CAC has promulgated over fifteen policy documents on the regulation of a variety of online content production services, from social media to search engines, mobile applications (APP), and online news production. Apart from demanding that all content production and distribution adhere to the law and official ideology, these documents also specify regulations on employee management and user registration, as well as punitive measures for any breaches of these regulations. According to the requirements, platform companies are fully responsible for all content circulated and will be "interviewed" (约谈 *yuetan*)—the code word for this in China is being invited for tea—when any of it violates the law or regulations. For example, in April 2018, Kuaishou and Toutiao were both "invited for tea" by the CAC for "ignorance of the law and disseminating programs that are against social moral values."[37] The CAC required the two companies to effect a "comprehensive rectification." As a result, their websites and apps shut down thousands of user accounts, including Tian You's, for posting "unhealthy content" and set up special official accounts for disseminating "positive and healthy values."

Thus, under the policy agenda of "Internet+" and "Mass Entrepreneurship and Innovation," the state's aspiration of economic restructuring drives but also shapes the platformization of Chinese cultural production. The state–corporate relationship is largely complicated due to the state's dual concern with economic restructuring and cultural regulation and social stability. Just as Sezgin and Binark show in their chapter on the Turkish digital game industry, this state–commerce relationship renders Kuaishou's content production acutely contingent and, we argue, distinguishes the platformization of cultural production in contemporary China from that in the West and constitutes a third dimension to what Nieborg and Poell summarize as the "contingency" of platform cultural production.[38] According to them, this contingency consists of "platform dependency" and "contingent commodities." The former refers to the dominant power of only a few platforms, such as Google, Apple, Facebook, Amazon, and Microsoft (GAFAM) in the West, and Baidu, Alibaba, and Tencent (BAT) in China, which "allow[s] content developers to systematically track and profile the activities and preferences of billions of users."[39] The latter refers to how platforms' content commodities are made continually "malleable, modular in design, and informed by datafied user feedback, open to constant revision and recirculation."[40] The power of the state, in the case of China, engenders a third dimension of contingency that constantly shapes the practice of cultural production on Chinese platforms: the platform commodities have to constantly adapt to not only datafied user practice but also state sanctions and regulations. But how is this contingency further translated in the digital affordances of Kuaishou? How does this networked platform governance affect the creator subjectivity and culture produced on the platform? Will it also lead to conflicted experiences among content creators, as shared by Turkish game developers studied by Sezgin and Binark? The following two sections will address these questions.

Walking through Kuaishou: Algorithmic and Digital Governance

Under the algorithmic logic, Kuaishou forges an ostensibly decentralizing and democratic system for content production and selection. In principle, everyone is treated equally by the algorithmic machine, whether they are a movie star or a migrant worker. The key for content production is to obtain as much online traffic through the creative content as possible. According to the online archives of the Kuaishou webpage[41] and its update records in Apple's app store, Kuaishou has described itself through slogans like "something interesting" and "record the world, record you" (2016–2018). On its webpage, three sentences appear under this new slogan: "Discover a real but interesting world. Be loyal to the self while not feeling lonely. The same town with the same mood." Such lines conjure up ideas of worlding, of the self, and of locality, thus grounding the contents offered in the everyday realities of China. The images shared on the webpage and in the app store further strengthen that sense of everydayness: ordinary young Chinese are captured on everyday occasions, while traveling, at home, with pets or babies, and so on. Keywords like "real," "self," "interesting," and "same," together with the photos, are indicative of the vision of Kuaishou: to invite "grassroots individuals" to discover and share the interesting moments in their own and others' "real" everyday lives (Figure 1). By promising an interesting, real, individualized, but not lonely online community, Kuaishou absorbs users' creativity into its platform economy.

Kuaishou is a free app, and its revenue sources consist mainly of in-app advertising and a gifting economy through live-streaming. As a typical content platform that connects multisided markets, Kuaishou offers two ways for advertising. The first one is called "fans headline" (粉丝头条 *fensi toutiao*), which allows content producers to promote their video content on the platform. According to the app's description, by paying RMB37.9 a posted video can gain an increase of 10,000 views from end-users. Producers can simply click on the "fans headline" button under the settings menu of the app interface. Another form of advertising is offered to third-party companies or brands that intend to buy advertising space on the interface. Commercials are mixed with user-generated videos and fed to targeted viewers by the algorithm. Kuaishou has not publicly specified the cost of its advertising

FIGURE 1. Vision of Kuaishou (SCREENSHOT FROM KUAISHOU'S OFFICIAL WEBSITE)

space, but a new media agency discloses that, apart from the one-off service fee of RMB5,600, advertisers pay RMB0.2 for each click.[42] Another important revenue source comes from the gifting economy in Kuaishou's live-streaming service. Only a select group of users are authorized to live-stream on the platform. Streamers interact with their fans during the show, and fans use *kuaibi* (快币), a virtual currency on Kuaishou, to buy and send virtual gifts to their favorite streamers. An amount of RMB1 can buy 10 *kuaibi* and the price of a virtual gift varies from 1 to 188 *kuaibi*. According to the platform's regulations, after deducting 20 percent for tax, the platform company earns half of the gifting income while streamers only obtain less than 40 percent. Clearly, Kuaishou's business model is largely dependent on how much data and data traffic the platform can collect from users—the raw material for the capitalist production in the platform age (see also introduction). The more popular its contents are, the higher financial returns the platform and its complementors can achieve.

Kuaishou has a simple interface. One can use email or social media accounts such as Wechat, Weibo, Facebook, or Google for registration.[43] There are three tabs on the main interface: "following" (关注 *guanzhu*), "trending" (发现 *faxian*), and "featured" (精选 *jingxuan*) (Figure 2). The default tab is "trending," which lists videos selected and pushed by the recommendation algorithm. There is no category selection button under the tab, and videos appearing here seem to be randomly selected. After using the app for a while, the streaming list will be updated and fed with new contents that are further calculated by the algorithm. Most of these videos are "trending": the majority of them have obtained at least hundreds of likes and most of them were posted that day. Usually one of these videos listed is promoted content. The number of videos and their genres will increase as the app is used over time. Under the "following" tab, content is listed in chronological order from the accounts followed by the user. The recommendation algorithm is not applied here since the user's preferences are quite clear. For content producers, this tab provides a window to interact directly and continuously with their target audience. The "featured" tab arranges videos in a similar way as TikTok, and users need to scroll down to see next videos. The platform gives priority in this tab to accounts that have bought "fans headline" services and accounts that are live-streaming. By adding a geolocational feature to the streaming system, the platform takes the opportunity of exploiting users' offline real-life social networks, which might create more user engagement.

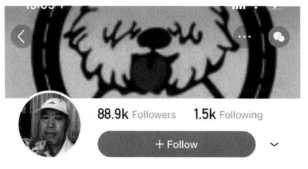

FIGURE 2. Screenshot of Kuaishou interface

To post content on Kuaishou, users do not need to change to a different account. By simply clicking on the camera icon above the interface, they can upload and post a short video up to 57 seconds long. Users can use smartphones to capture real-time moments and edit them with background music or animation effects provided by the app. They can also use the app to publicize premade, more professional content. As with other Chinese internet service providers, a mobile phone number is needed for verification. Before they can be streamed, all videos uploaded to the database will be analyzed by the algorithmic system. According to the three algorithm engineers and computer scientists interviewed, the algorithmic recommendation system of Chinese platforms has four basic components: content analysis, user analysis, evaluation, and security auditing. The first two use computation models to analyze and classify various content and user data. Based on this datafication, diverse content is tagged and distributed automatically among users who are predicted by the algorithm as target groups. The evaluation component fixes and optimizes the recommendation system based on the feedback from its previous operation. Finally, the security auditing component checks, filters, and censors all kinds of online content, including that provided by content producers and interactive content such as end-user comments. Through AI machine learning, the auditing system will achieve increasing accuracy. This is only a simplistic summary of the algorithm recommendation system. The technical components are very complicated and require enormous financial investment. For security auditing systems in particular, AI is not yet safe enough, meaning that Chinese platform companies often hire manual teams for online censorship. The state's requirement of a "positive," "healthy" internet culture thus increases the operational costs for these platforms. For instance, one of Kuaishou's human resource managers told us that it recently recruited three thousand new employees in branches in Harbin, Chengdu, Yancheng, Tianjin, and Wuhan to conduct manual censorship and online surveillance.

Verified "legal" content will then be pushed to a small group of end-users for the first-round feeding—including geographically nearby users, subscribers, and those predicted by the AI system as "potentially interested users." After the first twenty-four hours, the system will evaluate the content based on the feedback of user interactive data and decide whether the content is worthy of a second- or third-round push. If one buys the "fans headline" service, the posted video will receive the purchased amount of data traffic. As the platform's most valuable asset, the data traffic generated by the content will bring subscriptions, high interactivity, and subsequent advertising opportunities for the account holder. If their account remains highly active for weeks, achieves a large number of subscriptions, and has no history of violating the user regulations of the platform, content producers can contact the customer service for authorization of a live-streaming function. According to the platform's regulations, there are three standards for the evaluation of such an application: the account should have a high interactivity, including continuous uploading of original content with a large number of followers and online interactions; the account adds to the positive image of the platform and does not have any record of violating the regulations; and users should link their account to a mobile phone number to eliminate the risk of being hacked. In 2017 and early 2018, Kuaishou stirred discussion among the public and was "interviewed" by the CAC due to some online hosts' "vulgar" and "unhealthy" behaviors during live-streaming. As a result, the platform has become more cautious in regulating the live-streaming service. The ambiguous rhetoric of the regulations gives it ample leeway and power to control and manage the live-streaming service in accordance with its own interests and those of the state.

The algorithmic system and its immense database remain largely invisible to content producers, and Kuaishou holds a powerful position in its platform system, similar to that of Instagram. The company never discloses any technical details of its algorithms to users. In addition, content producers only have very limited access to interactive data, such as the number of followers, comments, and likes, through the app's user interface. They can check their followers' public homepage, but detailed user data such as demographic and behavior information are not available. Apart from restrictions on illegal or malicious content that violates Chinese law, the Kuaishou user agreement also prohibits users from any unauthorized commercial activities such as selling products or private advertisements. By posting any content on the platform, users automatically grant the Kuaishou company a "world-wide, royalty-free, non-exclusive, sub-licensable and perpetual (unless withdrawn expressly by you) license, to use the content uploaded (including but not limited to copy, publish, release, as well as adapt, reproduce, translate, transmit, perform and display in original form or other forms)."[44]

This user agreement patently panders to the company's commercial interests and to the Chinese state's requirement of a conforming and "positive" culture. Any violation would lead to punishments such as changing and deleting posted content or suspending and terminating the user's ability to use Kuaishou.

At a time when the state government is tightening its control over the cultural and media sectors, Kuaishou and other platforms will also reinforce their management of online content. This underlines the unequal relationship between creators and the platform company. Users engage with Kuaishou by encountering a wide range of options that create the semblance of choice, agency, and individuality, yet the Party-approved message framework and the profit-driven AI algorithms indicate that what happens on the site is constrained by corporate and Party directives—we get the look of freedom within a heavily structured and surveilled network. The governance of digital platforms creates a pervasive sense of uncertainty and insecurity among content producers. In the face of this, many content creators feel obliged to post some sentences on their homepage expressing their appreciation of Kuaishou, such as "thank you Kuaishou for providing such a wonderful platform," "I support Kuaishou for transmitting positive value," and so on (Figure 3).

Our analysis so far shows that the algorithm-based system of Kuaishou epitomizes the platform contingency caused by the special state–platform relationship, which, as indicated earlier, distinguishes the Chinese platformization of cultural production from that in the West. As a result, platform governance is subject to state regulatory power, both promoting and circumscribing the platformed cultural production. The algorithmic machine allows Kuaishou to achieve a maximal incorporation of creativity from "grassroots individuals," but state–platform contingency also sets limits to the automation of platform governance—for

FIGURE 3. Screenshot of one user's homepage

example, in requiring manual censorship. But how do creators then deal with this contingency and the need to navigate the tightrope between censorship and creativity, between production and commodification, between subjectivity and being subjected?

Unlikely Makers, Unlikely Aesthetics?

Wang Qian grew up in the countryside of Dazhou, Sichuan, a province in west China. At the age of fifteen, after graduating from middle school, he followed his relatives and went to Shenzhen, becoming a factory worker. In 2016, inspired by the stories of people making money through Kuaishou, he quit his job at the factory and decided to make videos on Kuaishou. Using the name "brother Qian"

(谦哥 *qiange*), he performs and teaches magic tricks in videos and live-streams on Kuaishou. Wearing a stylish hat and facial makeup, Wang Qian looks and performs like a professional magician in his videos. Yet he never received any magic training, and all his skills were learned from the internet. After two years, his account had over 1 million subscriptions. This large fan base allows him to sell and advertise magic props through the platform. As Wang Qian disclosed in our interview, the digital business enabled by Kuaishou generates an average sale of 70,000 RMB per month. Deducting production costs and salaries for his assistants, Wang's monthly income can reach 50,000 RMB (6,000 euros). Compared to his job at the factory, his new job has not only multiplied his earnings, but has also changed his appearance and identity: he seems to have gotten rid of the stigmatizing label of "migrant worker" to become part of the affluent, fashionable, and popular "online celebrity" (网红 *wang hong*) class.[45] Through Kuaishou, then, Wang has jumped from the "sweatshop" in Shenzhen into the urban creative class.

Wang Qian's experience is not uncommon on Kuaishou. In Li et al.'s study of the use of Kuaishou among a group of rural students, the video-sharing app is seen to allow these low-income rural youths a way to "express their resistance against education" through the circulation and production of the *"shehui ren"* (社会人, society man) subculture.[46] As it did for Wang Qian, Kuaishou promises these rural youths an upward socioeconomic mobility through capitalizing on their memories and creativity in the production of the *"shehui ren"* subculture.

Its massive popularity among the Chinese rural population, and the produced culture and aesthetics significantly distinguish Kuaishou from other Chinese and Western social media platforms. Most Kuaishou users are from the urban lower social class and young people from rural society, enabled by Kuaishou to "record the world and themselves." According to a manager from Kuaishou, the company has never tried to sign or promote any particular "online celebrity." Instead, the platform embraces an aesthetics of the vernacular, which can be described as foregrounding

> the un-hip, the un-cool, and possibly the downright square, [it] embraces those marginal and non-glamorous creative practices excluded from arts- and culture-based regeneration. Vernacular forms of creativity are neither extraordinary not spectacular... but are part of a range of mundane, intensely social practices.[47]

This vernacular aesthetics evidenced in the online culture of Kuaishou seems to align with the countercultural values as illustrated in Mutibwa and Xia's analysis of the Chinese Maker Movement in Shenzhen. To further explicate this vernacular aesthetics circulated on Kuaishou, we selected and analyzed 200 trending videos and the everyday user activities of 20 popular Kuaishou accounts. These selected short videos constitute what Lauren Berlant has called a "silly archive," which may be "the silliest, most banal, and . . . of erratic logic" in the everyday experiences of ordinary citizens.[48] It is precisely its "very improvisatory ephemerality," its "very popularity," and "its effects" on everyday life, according to Berlant, that makes such a silly archive worthy of serious reading.[49]

In our selection of videos, we observed five recurring genres of content:

1. Everyday life: cooking, cosmetics, pets, family life, etc.
2. Country life: fishing, hunting, crafts, vernacular landscape, etc.
3. Creative skills: singing, magic, dancing, fitness, professional skills, etc.
4. Fiction micro film
5. "Positive value content" mostly produced by official sponsored accounts.

For the first four genres, the idea of "grassrootsness" is crucial; it is performed to add "authenticity" to the videos, to make them look more real and closer to the audience's own life. For Wang Qian, behind his polished appearance in the videos, such grassrootsness is demonstrated by his accent and his way of performing magic. Unlike professional magicians, he shows only forms of magic that he learned from the internet. More than that, he also unveils and teaches magic to his fans. In one of his videos, he remarks at the end: "Come on brothers. With this trick you will find a girlfriend!"

Other video makers also choose to deliberately display their underclass identity through their accent, dress, skin color, or behavior. For instance, in a series of videos showing cosmetic skills, the female model has quite dark skin and chubby cheeks, which does not meet the current standard in China for a "beautiful girl." Thanks to the skillful use of cosmetic techniques and the special products used, the model has her appearance drastically changed, with fairer skin and thinner cheeks. Her new look (Figure 4) is still not comparable to that of professional models in television advertisements, yet videos like these are quite popular on Kuaishou and within a few hours can easily gain hundreds and thousands of likes from users. The secret to the high popularity of these videos is precisely the "grassrootsness" and "authenticity" they aim to represent: not every ordinary person is born with the beauty of a movie star, yet, by virtue of the "right" makeup and techniques, video makers convince their audience that they, too, can change their imperfect physical appearance. On the homepage of another account named "Zhang Deshuai," the

FIGURE 4. Before and after using makeup (SCREENSHOT OF ONE SELECTED VIDEO)

video maker identifies himself as a "country lad" (农村小伙 *nongcun xiaohuo*) and posts homemade micro-films. These films usually choose shabby villages as a background and tell amusing stories about relationships, family life, friendship, etc. However, in contrast to the rural landscape shown, the cast members in these videos always dress in trendy fashion with a stylish haircut, while the hilarious story lines are not necessarily about "country life." The characters in the films, for example, talk about "watching movies," "shopping," "drinking milk tea," and "buying a car." Thus, while these films choose rurality as their background, in terms of the characters they show and the stories they tell, they push the limits of rurality and intentionally parody the trendy life of Chinese urban youth. From Wang Qian's magic demonstrations and the popular cosmetic videos to these self-made fiction films, the aesthetics of the videos on Kuaishou articulate the imagination of Chinese "grassroots individuals" who are marginalized in mainstream popular culture. It is this imagination from the marginal that fascinates many Kuaishou users, who to some extent experience this grassrootsness and marginalization in their own everyday lives.

At the same time, as we pointed out in the previous section, content production on Kuaishou is not immune to state surveillance. Since being interviewed by the CAC in April 2018, "positive and healthy values" guide content regulation on the platform. The once very popular "crazy videos," such as those depicting adolescent pregnancy and self-abuse, have been banned and deleted. The platform has also established a new genre of "positive-value" content. Apart from its own official account "Kuaishou positive value" (快手正能量 *Kuaishou zheng nengliang*), the platform invites government institutions such as public security bureaus to open accounts and post videos on "everyday ethical models," "Chinese economic achievements," "the positive image of soldiers and the police," "the official policy and ideology," etc. The algorithmic system has been set to support the videos uploaded by these accounts, which is why, in August 2018, they featured 7 out of the 10 most viewed videos on Kuaishou.[50]

In this regard, content producers on Kuaishou have to meticulously calculate their creativity, to remain in line with the platform's so-called "value orientation" while also making their content attractive to the online audience. Once again, dovetailing with the argument made in the introduction, the design of Kuaishou's algorithmic system and its promised inclusiveness falls into the state top-down governance. Moreover, for these creative individuals, the platform and its digital affordances denote not only a way of performing creativity but also an effective tool for making money and building a career. Data traffic becomes a crucial asset that every creative producer aspires to accumulate in as high a quantity as possible. To do so, they first need to understand and utilize the various digital affordances of the platforms. They should, for example, update their accounts on an everyday basis. From their profile photo to their user name, everything that can give end-users a sense of what the account is about needs to be deliberately designed and optimized. To be creative through the digital, one has to know what, how, and when to create, and for whom. The constant posting and streaming also requires good time-management skills. On Kuaishou, producers normally choose to post their videos in the evening around 8 pm, a time when most high school students, one of the largest user groups on Kuaishou, are at home and have just finished their homework. What Melissa Gregg identifies as the "presence bleed"— how digital and communication technologies enable "work to invade places and times that were once less susceptible to its presence"—becomes imperative for platform-based creative work.[51] The aim of this intensified and extensified[52] work for content producers is to generate profits, which also leads to users' appropriation of the digital technologies for their own business purposes.[53] Although Kuaishou prohibits unauthorized advertising and commercial activities, video makers can still find their own ways to avoid the platform's supervision. Some streamers integrate contextual advertisements for third-party merchants in their short video and live-streaming performances. For example, someone posts videos of their pets on Kuaishou and lists their Wechat account number on the homepage to sell pet food, using acronyms such as "WX" or icons like "V ♥" as a substitute for Wechat (微信 *weixin*) to dodge the platform's AI monitoring (Figure 3).

The high interactivity of the digital platform requires content producers not only to strategize their creativity for business purposes, but also to manage their affects and personality to cultivate intimacy with their target users and audience. On Kuaishou, a phrase that appears frequently in short videos is "Come on bro! Double tap 666! Follow me." The action of double tapping on a video equals a "like" from a viewer, and "666" in Mandarin is homophonic to *liu* (溜), meaning "cool" or "awesome." These words are often spoken in a euphoric tone with local accents. The aim is to add a sense of authenticity to the videos and develop intimacy with the audience. To gain more popularity and subscriptions, one of the strategies used by content producers is to set up a special *"renshe"* (人设, character) to perform a certain personality through various creative practices that will affect and create intimacy with viewers, who will later become their followers, or fans. As is exemplified by the videos analyzed above, on Kuaishou, a personality is carefully nurtured and maintained through performing "grassroots authenticity." A frequent discourse that emerges out of these diverse stylizations is that of being "real-life" and *"jiediqi"* (接地气, down to earth), underlining how the personalities created should be relevant, if not identical, to those of the platform's users.

As a result, on Kuaishou, platformed cultural production is entangled with the production of affects. These affects, such as "a feeling of ease, well-being, satisfaction, excitement or passion," are produced through the labor process of platformization, "expressing a certain state of body along with a certain mode of thinking."[54] By exploiting the various digital affordances and *"renshe,"* "grassroots" content producers have, on the one hand, become self-employed creative entrepreneurs for whom creativity, life, and individuality are constantly calculated according to the accounting of costs and profits. On the other hand, in the everyday production and management of affect through the digital affordances provided by Kuaishou, these creative individuals also become aspirational creative workers motivated by the platform's "promise of social and economic capital; yet the reward system for these aspirants is highly uneven."[55] A data-driven economy becomes the common model that drives all the parts becoming complementors of the platform. Data and datafication not only matters for giant corporations and institutions,[56] but they also become crucial production tools and assets for these new, "unlikely" creative subjects on Kuaishou.

Conclusion

This chapter has studied a special group of creative workers—the content creators on the Chinese social media platform Kuaishou, enabled by the emerging Chinese platform creative economy. We take Kuaishou as a case of multipolar platformization. We first examined the state-platform contingency caused by the complicated relationship between Kuaishou and the state governance of culture and economy, and how such contingency is embedded in the digital algorithmic system of the platform. This third dimension of platform contingency distinguishes the functioning ecology of Chinese media platforms from those in the West. This contingency maximizes the subsumption of individual vernacular creativity in China's platform creative economy, while also enabling marginalized "grassroots" Chinese to become "unlikely" creative workers. These affordances of the platform inspire us to steer away from a focus on risk and uncertainty in the disruption/structure paradox, just as they guide us away from the urban that dominates creative labor studies.

At the same time, through this production of an "unlikely" creative class, the platformization of cultural production accommodates the Chinese state's "entrepreneurial solutionism,"[57] which, exemplified by the state's policy on Internet+ and Mass Entrepreneurship, takes digital technology and entrepreneurship as the solution to China's social, economic, and cultural problems. The platform economy thus provides opportunities for "grassroots individuals" from diverse backgrounds to become creative workers, pandering to the state's goal of restructuring the economy.

Importantly, this grassroots digital entrepreneurship has also transcended the passive "digital labor" and "prosumer" models some critical politic economists have identified.[58] Despite the institutional regulation and censorship of the internet, these grassroots creators actively participate in the Chinese platform creative economy, appropriating the algorithmic digital system and negotiating with the state/platform governance to achieve their own creative and financial aims. Within their experiences of creation and monetization, we can find moments of play, if not resistance—moments in which the official narrative of the "China Dream" is juxtaposed to multiple dreams from actors that hardly ever get a face or a voice in Chinese mainstream media.

NOTES

This is a rewritten version of an article that appeared with the same title in *Social Media + Society*, November 2019, https://doi.org/10.1177/2056305119883430. This study is part of the project ChinaCreative, funded by the European Research Council (ERC consolidator grant no. 616882).

1. Jiu Xing, "Lonely Hero," 2016, http://bbs.tianya.cn/post-641-31626-1.shtml.
2. In rapper culture, MC stands for "Microphone Controller" or "Master of Ceremonies," often used as a title for skilled rappers. Tianyou chooses this title to identify himself as a Chinese rapper. The appropriation of English terms like MC is indicative of the imagined origins of rap music—the United States—which, as de Kloet has argued elsewhere, sparks off a strong localizing aspiration in which Chineseness as well as specific localities (e.g., Wuhan, Shanghai) are being articulated; see Jeroen de Kloet, *China with a Cut—Globalisation, Urban Youth, and Popular Music* (Amsterdam: Amsterdam University Press, 2010).
3. *Hanmai* literally means "shouting with microphone." Chinese online rappers like Tianyou created and performed it. These self-claimed MCs shout out rhythmed lyrics, usually rephrased in classical Chinese with some popular online slangs, in high beat music (usually downloaded from the internet).
4. David Craig, Jian Lin, and Stuart Cunningham, *Wanghong as Social Media Entertainment in China* (New York: Palgrave Macmillan, 2021); Junyi Lv and David Craig, "Firewalls and Walled Gardens: The Interplatformization of China's Wanghong Industry," in *Engaging Social Media in China: Platforms, Publics, and Production*, ed. Yang Guobin and Wang Wei (East Lansing: Michigan State University Press, 2021), 51–74.
5. H. Arcbering, "Hanmai, a Chinese Rap-like Performance, Symbolizes a Cultural Interest in Rural China," January 21, 2017, https://medium.com/@arcbering/hanmai-a-chinese-rap-like-performance-symbolizes-a-cultural-interest-in-rural-china-79c55fce3108;

Javier C. Hernández, "Ranting and Rapping Online in China and Raking in Millions," *New York Times*, September 15, 2017.

6. Tricia Rose, *Black Noise: Rap Music and Black Culture in Contemporary America* (Hanover, NH: Wesleyan University Press, 1994).

7. Xiaoli Chen, "Live-Streaming Star Banned over 'Offensive' Content," *Shine*, February 13, 2018, https://www.shine.cn/news/nation/1802130400/.

8. The Cyberspace Administration of China, "The White Paper on Chinese Digital Economy [中国数字经济发展白皮书]," 2017, 19, http://www.cac.gov.cn/files/pdf/baipishu/shuzijingjifazhan.pdf.

9. Patrick Shaou-Whea Dodge, ed., *Communication Convergence in Contemporary China: International Perspectives on Politics, Platforms, and Participation* (East Lansing: Michigan State University Press, 2020). Cyberspace Administration of China, "The White Paper on Chinese Digital Economy."

10. See chapter 1 of this book for a more detailed discussion on platform economy.

11. David Nieborg and Thomas Poell, "The Platformization of Cultural Production: Theorizing the Contingent Cultural Commodity," *New Media & Society* 20, no. 11 (2018): 4275–92.

12. P. David McIntyre and Arati Srinivasan, "Networks, Platforms, and Strategy: Emerging Views and Next Steps," *Strategic Management Journal* 38, no. 1 (2017): 141–60, https://doi.org/10.1002/smj.2596; Nieborg and Poell, "The Platformization of Cultural Production."

13. Danru Liu, "After Staying 5 Days in the Country-Side, I Finally Understand Why Kuaishou Can Attract 400 Million Chinese [过年在农村待了5天，我终于知道为什么快手能 横扫4亿中国人]," *36Kr.*, February 3, 2017, https://36kr.com/p/1721350995969.

14. Nieborg and Poell, "The Platformization of Cultural Production."

15. Nieborg and Poell, "The Platformization of Cultural Production."

16. Ben Light, Jean Burgess, and Stefanie Duguay, "The Walkthrough Method: An Approach to the Study of Apps," *New Media & Society* 20 (2016): 882, https://doi.org/10.1177/1461444816675438.

17. There are three ways we adopted for the selection of videos and accounts. To minimize any personal preference, we first use a new account to download the first 20 videos listed under the "trending" tab for 7 days. The second collection of videos are selected from the first 20 accounts on the ranking list of "the most popular live-streamers on Kuaishou," provided by xiaohulu.com, which is a third-party start-up company offering data analysis and operation service for content producers on major Chinese

content platforms. The final selection of videos consists of the 10 most-viewed videos on Kuaishou in August, data provided by "short video factory," another third-party company publishing business reports on Chinese short-video platforms.

18. Richard Florida, The Rise of the Creative Class: And How It's Transforming Work, Leisure, Community and Everyday Life (New York: Basic Books, 2002).
19. Richard Lloyd, Neo-Bohemia: Art and Commerce in the Postindustrial City (London: Routledge, 2005).
20. Niels van Doorn, "Platform Labour: On the Gendered and Racialized Exploitation of Low-Income Service Work in the 'On-Demand' Economy," *Information, Communication & Society* 20, no. 6 (2017): 898–914.
21. George Ritzer and Nathan Jurgenson, "Production, Consumption, Prosumption: The Nature of Capitalism in the Age of the Digital 'Prosumer,'" *Journal of Consumer Culture* 10, no. 1 (2010): 13–36.
22. Nieborg and Poell, "The Platformization of Cultural Production," 4.
23. Yu Hong, *Networking China: The Digital Transformation of the Chinese Economy* (Urbana: University of Illinois Press, 2017), 10–13.
24. Internet+ is not the same as "Made in China 2025," which is a different national strategy aimed at uplifting China's manufacturing industry through technological innovation; Internet+, however, focuses on digital infrastructure and platformization.
25. The State Council, "Instructions on Promoting 'Internet+' [国务院关于积极推进'互联网+'行动的指导意见]," 2015.
26. The State Council, "Instructions on Constructing Platforms to Promote Mass Entrepreneurship [国务院关于加快构建大众 创业万众创新支撑平台的指导意见]," 2015; Yu Hong, "Pivot to Internet Plus: Molding China's Digital Economy for Economic Restructuring?" *International Journal of Communication* 11, no. 21 (2017): 1486–506.
27. State Council, "Instructions on Constructing Platforms to Promote Mass Entrepreneurship."
28. Keqiang Li, "Internet+ Has Provided a Wide Stage for Mass Entrepreneurship ['互联网+'为大众创业、万众创新提 供了广阔的舞台]," *People.Com*, March 20, 2018, http://lianghui.people.com.cn/2018npc/n1/2018/0320/c418651-29878641.html.
29. All of them are private companies and receive financial investment from Chinese internet giants Baidu, Tencent, and Alibaba. The headquarters of Kuaishou Company is located in Beijing and receives investment from Baidu, Tencent, and several other venture capital firms.
30. The notion of "grassroots individuals" (草根 caogen) resonates with the often used

"common people" (老百姓 laobaixin). While we use the term in quotation marks for its prevalence in Chinese discourses, we are aware that these are highly problematic terms that produce a binary division in society between the people and the elite, ignoring further stratifications and more subtle class differences, and hence our use of quotes when using the term.

31. Tarleton Gillespie, "The Relevance of Algorithms," in *Media Technologies: Essays on Communication, Materiality, and Society*, ed. Tarleton Gillespie, Pablo J. Boczkowski, and Kirsten A. Foot (MIT Press Scholarship Online, 2014), 192.
32. We will come back to examine the technical and algorithm system of Kuaishou in the following section.
33. Qiming Huo, "The Cruel Bottom Class: Chinese Rural Society in a Video App [残酷底层物语：一个视频软件的中国农村]," *Tencent News*, June 8, 2016, https://xw.qq.com/news/20160609003283/NEW2016060900328301.
34. Jean Burgess, "Hearing Ordinary Voices: Cultural Studies, Vernacular Creativity and Digital Storytelling," *Continuum: Journal of Media & Cultural Studies* 20, no. 2 (2006): 201–14.
35. Wanning Sun, "Mission Impossible? Soft Power, Communication Capacity, and the Globalization of Chinese Media," *International Journal of Communication* 4 (2010): 66.
36. The Cyberspace Administration of China, "Detailed Regulations on the Administration of Internet News Service [互联网新闻 信息服务许可管理实施细则]," 2017.
37. Jing Liu, "Kuaishou and Toutiao Got Interview! [快手、今日头条、火山小视频被约谈]," *People's Daily 2*, April 8, 2018.
38. Nieborg and Poell, "The Platformization of Cultural Production," 2.
39. Nieborg and Poell, "The Platformization of Cultural Production," 2.
40. Nieborg and Poell, "The Platformization of Cultural Production," 2.
41. See https://www.Kuaishou.com/.
42. Zhiyuan Advertising, "Introduction to Kuaishou Platform [快手平台简介]," 2018, https://www.zdzynet.com/fwcp/spyx/13.html. Kuaishou does not sell advertisement of financial or medical products, or other social media platforms.
43. It is quite ironic that a China-based platform includes both Facebook and Google in its registration interface; this shows how the censorship toward both is anything but clear-cut or univocal.
44. Kuaishou, "User Agreement," n.d., https://www.kuaishou.com/about/policy.
45. See Craig, Lin, and Cunningham, Wanghong as Social Media Entertainment in China.
46. Miao Li, Chris K. K. Tan, and Yuting Yang, "Shehui Ren: Cultural Production and Rural Youths' Use of the Kuaishou Video-Sharing App in Eastern China," *Information,*

  Communication & Society 23, no. 10 (2020): 1499–514.
47. Tim Edensor et al., "Introduction: Rethinking Creativity: Critiquing the Creative Class Thesis," in *Spaces of Vernacular Creativity: Rethinking the Cultural Economy*, ed. Tim Edensor et al. (New York: Routledge, 2009), 10; Burgess, "Hearing Ordinary Voices."
48. Lauren Berlant, *The Queen of America Goes to Washington City* (Durham, NC: Duke University Press, 1997), 12.
49. Berlant, *The Queen of America Goes to Washington City*, 12.
50. Short Video Factory, "TOP 20 List of the Most Viewed Short Videos in August 2018," 2018, http://www.sohu.com/a/252570987_820218.
51. Melissa Gregg, "Spousebusting: Intimacy, Adultery, and Surveillance Technology," *Surveillance and Society* 11, no. 3 (2013): 2, https://doi.org/10.24908/ss.v11i3.4514.
52. According to Jarvis and Pratt, contemporary media and cultural industries give rise to an increasing extensification of work, referring to the distribution or exporting of work across divergent spaces/scales and times. Helen Jarvis and Andy C. Pratt, "Bringing It All Back Home: The Extensification and 'Overflowing' of Work: The Case of San Francisco's New Media Households," *Geoforum* 37, no. 3 (2006): 331–39.
53. Jarvis and Pratt, "Bringing It All Back Home."
54. Michael Hardt and Antonio Negri, *Multitude: War and Democracy in the Age of Empire* (New York: Penguin Books, 2004), 108.
55. Brooke Erin Duffy, "The Romance of Work: Gender and Aspirational Labour in the Digital Culture Industries," *International Journal of Cultural Studies* 19, no. 4 (2015): 441.
56. Jose van Dijck, "Datafication, Dataism and Dataveillance: Big Data between Scientific Paradigm and Ideology," *Surveillance & Society* 12, no. 2 (2014): 197–208, https://doi.org/10.24908/ss.v12i2.4776.
57. Michael Keane and Ying Chen, "Entrepreneurial Solutionism, Characteristic Cultural Industries and the Chinese Dream," *International Journal of Cultural Policy* 25, no. 6 (2019): 743–55.
58. Ritzer and Jurgenson, "Production, Consumption, Prosumption."

Technology Translations between China and Ghana

The Case of Low-End Phone Design

Miao Lu

As night falls in a small village in Kpone-Katamanso District, about 38 kilometers from Ghana's capital, Accra, only half of the households will be lit up. Because Kelly's house does not have electricity, she must turn on the flashlight of her phone to cook dinner. Kelly is using a Tecno feature phone that costs less than US$10. Its battery can last for three to five days after she charges it at her neighbor's house. While Kelly is preparing the dinner, her mother sits by the door, listening to the radio, an important means of both information and entertainment in the local community. In December 2019, I visited Kelly and her village with my local informant, Michael. This is an impoverished village where roads are unpaved, half of the households are unelectrified, television sets are rare, but mobile phones are ubiquitous. Except for some young people who have smartphones, most villagers are using cheap feature phones like Kelly's, the so-called "keypad phones" in Ghana or "dumbphones" in the West. In Kelly's village, the most popular phone brands are Tecno and itel, both from a Shenzhen-based company called Transsion Holdings (hereafter Transsion). Since setting foot in Africa in 2007, Transsion has surged from almost nowhere to become the largest mobile phone vendor there, capturing 52.5 percent of its market share in 2019.[1] In Africa, Transsion phones are known for their affordable prices and locally tailored features, including dual SIM cards, long battery life, and cameras optimized for darker skin tones.

In the business world, Kelly's village is an example of the "bottom of the pyramid" (BOP) markets where three billion people live on less than two dollars a day. Deemed as "unusable" by global capital,[2] the BOP population has long been left on the wrong side of the "digital divide." Much literature about these marginalized users centers on the theme of "everyday inventiveness," in which improvisation and innovation are used to cope with everyday breakdown and scarcity in the context of "pirate modernity."[3] In the tech world, humanitarian design emerges to engage with the BOP population through projects such as "One Laptop per Child," but has been questioned regarding its utilitarian framing, universalist assumption, top-down approach, and neglect of local differences.[4] By embracing so-called "inclusive capitalism," Silicon Valley companies are also initiating design projects to tap into the "fortune at the bottom of the pyramid."[5] Underlying such an approach is a type of neoliberal thinking that frames poverty as a product of market failure and identifies its solution as integrating the poor into the global market.[6]

In such contexts, why do Chinese tech companies such as Transsion make forays into Africa's BOP markets? How can they design products that are affordable and profitable in these markets? How do their practices differ from existing design approaches, whether corporate or humanitarian? What is their innovation? Will their design practice contribute to "technodiversity" by cultivating alternative social-technical worlds?

To address these questions, this study argues that design should be an important area of inquiry, both empirically and theoretically. Following Walter Mignolo's insights on the geopolitics of knowledge, this chapter maintains that a critical study of design should interrogate *who* designs for *whom* and from *where*.[7] A growing body of research has been problematizing the normative view of Western design, which is largely white, male, and urban-oriented.[8] Why is design so "white"? What can we learn from non-Western designers?

Where we design matters. To provincialize Western design practice and acknowledge the geopolitics of design, the field of "design for/by/from the Global South" is thus a timely call.[9] Engaging with "the south of design"[10] is to reimagine technologies from a Southern perspective and explore alternative socio-technical worlds. As the largest mobile phone provider in Africa, Transsion raises critical questions about cross-cultural design for/by/from the Global South and complicates the binary logic of the North-South divide. In mainstream narratives about innovation, technologies are usually "designed" in the West (e.g., Silicon Valley) and then "diffused" to the rest of the world (e.g., Shenzhen). Is it possible for Shenzhen to be Silicon Valley's "South" and Africa's "North" at the same time? How will "multipolar" innovation play out in the China-Africa context?

Based on fieldwork in China and Ghana, this chapter examines the dynamics and tensions of designing for Africa's BOP users through the case study of Transsion's low-end phone brand itel. By tracing the origin and transformation of design practices in Silicon Valley and Shenzhen, this study presents a nuanced understanding of design thinking and doing in different cultures. Using "technology translation" as an analytical framework, this chapter explains itel's multi-actor design network and elaborates on various types of technology translations by considering the cases of SIM cards, battery, and camera, three features that have been viewed as critical to Transsion's success in Africa.[11]

Moreover, while the Trump administration's "America First" policy emphasizes isolationism and leaves the African markets undervalued, if not untapped, China remains a firm advocate of globalization and multilateralism.[12] From the "Going-out" policy in 2001 to the recent "Digital Silk Road,"[13] China has initiated various state-led projects and issued favorable policies to encourage its enterprises to invest overseas. While Chinese state companies often engage in controversial, large-scale infrastructure projects in Africa, Transsion attempts to bring mobile devices to often-neglected rural markets and address local needs for digital technologies through design and innovation. Thus, Transsion as a case not only provides an opportunity to explore alternative modes of development and innovation but also sheds new light on the trajectory of China-led globalization and its potential impact on the geopolitics of information.

Silicon Valley: From Design to Design Thinking

Silicon Valley tech giants, most notably Apple, often self-identify as "design-driven" companies. In his book, Barry Katz traced how an ecosystem of engineers, designers, sociologists, and anthropologists, as well as market researchers, lawyers, and venture capitalists, gradually took shape in Silicon Valley, transforming it from a provincial outpost to an elite design center.[14] Deeply rooted in this ecosystem, Silicon Valley design is often celebrated as research-based, innovation-driven, and human-centered. In particular, Apple's Steve Jobs accorded to design a place seldom seen in previous tech companies. Jobs emphasized not only the product design but also the design of the company and the image it sought to project; it was "a complete design mindset, a way of thinking and making sense of things."[15]

In the past decade, the term design thinking has gained popularity in the business world. Design thinking is the idea that design, as a creative way of problem-solving, can be applied to the totality of life, which represents a radical expansion of the meaning of design.[16] As Canadian designer Bruce Mau claims, "It is not about the world of design; it's about the design of the world."[17] Underlying this claim is a cultural imaginary that places design as the main force and designers as the main agents of transformative change. With little doubt, Silicon Valley is the epicenter of change and future-making.

This phenomenon has drawn criticisms from several scholars. Kimbell traced three accounts of design thinking: as a cognitive style, as a general theory of design, and as a resource for organizations.[18] According to her, the adoption of this term in managerial discourse, particularly business schools, is a way of depoliticizing managerial practice so as to balance organizational tensions between exploration and exploitation. Adopting the labor perspective, Irani argued that Silicon Valley's shift to design thinking was an effort to defend American distinctiveness in response to the competition from Chinese designers.[19] For a long time, Asia has been viewed as a source of cheap labor and criminal pirates; in contrast, Silicon Valley, which is at the apex of the global labor hierarchy, is a place for the world's most creative jobs. Quickly moving beyond the copycat era in recent years, Chinese designers have been offering design services that are often more efficient and less expensive. Driven by the anxieties about global labor competition and the changing economic order, Silicon Valley shifted from design to design thinking. Hence, "design thinking" was constructed as a form of expertise superior to the craft skills of designers and "workers" in developing countries.

Shenzhen: From *Shanzhai* to Indigenous Innovation

In the global information and communication technology (ICT) industry, the cases of Apple and its contract manufacturer Foxconn are widely used to criticize the structural inequality of the world system.[20] Economically, with global tech giants monopolizing core technologies, manufacturers in Shenzhen such as Foxconn can claim only a tiny portion of the profit. Discursively, to maintain existing global division of labor, Shenzhen as a site of design is much less legitimate than it is as a site of manufacturing. However, the cultural logic of manufacturing in Shenzhen does not always obey the authorized track but results in excessive technological mimesis.[21] Hence, Shenzhen's design culture is often dismissed as copycat culture, with *shanzhai* phones as a notable example.

Shanzhai is a Chinese term used to describe the broad phenomenon of copycat productions in China. Recently, a growing body of literature has reflected on *shanzhai* as a form of disruptive innovation.[22] The *shanzhai* ecosystem, which is comprised of a web of interconnected actors, allows designers to react quickly and flexibly to the fast-changing market.[23] Technically speaking, the "turnkey solution" introduced by the Taiwanese company MediaTek avoids unnecessary patent issues for *shanzhai* makers and significantly reduces the barriers to entry for phone design.[24] Based on this solution, *shanzhai* makers can either adopt the common module design or add features to create new models. As Mutibwa and Xia have documented in this book, *shanzhai* culture often carries an open-source ethos and encourages grassroots design and innovation, offering new choices and possibilities to markets ignored by large multinationals.[25] In China, the proliferation of *shanzhai* phones was closely related to the rise of China's "information have-less" population,[26] as they could obtain important information and social support through these affordable phones.

Currently, *shanzhai* is no longer about copycat production in the informal economy but experiences a process of institutionalization.[27] Renaming and rebranding efforts from various actors also constitute a significant rhetorical shift in the *shanzhai* discourse. For example, the Shenzhen government has worked hard to reframe the city's image from a manufacturing hub to a city of design.[28] Chinese state media also attempt to link *shanzhai* to "indigenous innovation," a national strategy aimed to transform China into an innovative society.[29] In the 1980s and 1990s, Chinese ICT companies encountered great difficulty in acquiring foreign technologies, and thus, they were forced to build their own innovation capability.[30] For instance, China's efforts to develop its homegrown 3G standard were aimed to reduce dependence on foreign technologies, including the large patent fees paid to Western operators.[31] Through the critical lens of disruption/structure, these policies signify a techno-nationalist approach of mastering core technologies when latecomers attempt to (or have to) disrupt Western domination in digital technologies through homegrown innovation.[32]

The ICT industry is a landmark sector in China's pursuit of indigenous innovation. In 2007, China abandoned its license control over the manufacturing of mobile phones, which greatly enhanced the legitimacy of *shanzhai* phones.[33] Within a short period of time, some homegrown brands such as Huawei and Xiaomi have gained footholds in the global market. In 2017, eight of the top ten smartphone brands in China and three out of the top six smartphone brands worldwide were Chinese.[34] As exemplified by Huawei's Kirin processor,[35] Chinese phone makers have steadily moved up the global value chain.

To summarize, Shenzhen's transformation from a manufacturing center to an innovation hub has been shaped by several intertwined forces, ranging from top-down national policies to grassroots innovations like *shanzhai*. This demonstrates the heterogeneity of design culture and the innovative capacity of Southern countries. With a new generation of Shenzhen companies going overseas, they are reshaping regional, if not global, digital culture and landscape. In such contexts, the case of Transsion in Africa offers a fascinating glimpse of this ongoing transformation.

Framework and Research Methods

In science and technology studies, technology is often examined in relation to both its designers and users in a social-technical network. Inspired by the metaphor of technology as "text" or "script," the triadic model of "designer-technology-user" focuses on what is "written" into an artifact by designers (inscription) and how it is "read" by users (de-scription).[36] When users can only access pirated or secondhand technologies, they are completely disconnected from designers and the design process. Thus, researchers usually focus on the second part of the triadic model—"technology-user"—to examine users' appropriation and improvisation when engaging with these technologies.[37]

Figure 1 shows this study's proposed analytical framework, which adjusts the triadic model by introducing "translators" into the design network. "Translators" refer to people who translate technologies and related information for other actors. Through their labor of translation (Arrows 3 to 5), translators mediate the interactions between designers and users. Such an analytical framework suits this study for two reasons. First, as Transsion's target users, Africa's BOP population are using phones specifically designed for them instead of pirated or used ones. This means that the interactions between designers and users are not absent, as indicated by Arrow 6. Second, although not absent, it is still not easy for designers and users to directly interact with each other. Due to geographical, cultural, and

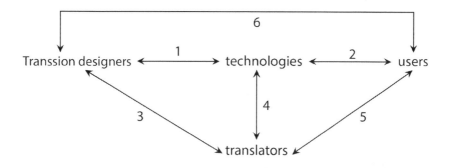

FIGURE 1. Proposed analytical framework

socioeconomic differences, various intermediaries are often indispensable. Hence, introducing translators to this network opens up a new space to understand the dynamics of designing technologies for low-end users.

This study focuses on Transsion's low-end brand itel. Transsion was founded in 2006 by former employees of Ningbo Bird, which was one of China's largest mobile-phone makers in the early 2000s. Transsion targets emerging markets like Africa and India as a way to avoid the fierce competition in China's domestic market. In Africa, Transsion sells mobile phones through its three brands: itel, Tecno, and infinix. Particularly, itel has a large market share and good reputation in Africa's rural markets. Moreover, deeply rooted in Shenzhen's maker ecosystem, itel emerges from, but goes beyond the early stage of the *shanzhai* model, providing an opportunity to explore the transformation of Shenzhen's design culture.

This study uses participant observation and semi-structured interviews as the primary research methods. From mid-July to mid-October 2018, I conducted participant observation at Transsion's Shenzhen headquarters.[38] Afterwards, I conducted semi-structured interviews with various Transsion employees. The second part of my fieldwork was conducted in Ghana from September to December 2019, where I visited itel shops, interviewed local dealers and users, and conducted observations in mobile-phone markets in cities (e.g., Accra and Kumasi), towns (e.g., Wenchi and Kintampo), and villages. All interviewees were informed that I was collecting data for academic research and their names would be anonymized in this study to protect their identities.

FIGURE 2. itel 5606 (PHOTOGRAPH © LU MIAO, REPRINTED BY PERMISSION)

Mapping the Design Network: Translating the Four "Ps" (Product, Price, Place, and People)

In Ghana, itel 5606 (see Figure 2) was one of the most popular feature phone models in 2019. Retailing at around US$10, itel 5606 has a flashlight, a wireless radio, a 2500mAh battery, dual-SIM cards, and even Facebook! When introducing their product positioning in Africa, product manager Z says that itel focuses on handsets under US$100: feature phones are usually retailed at three prices—US$8.00, US$15.00, and US$20.00—while smartphones are at US$50.00, US$75.00, and US$100.00, much lower than the global average.[39] How do itel designers decide the three "Ps"—product, price, and place? This is closely related to another "P"—people, itel's target users.

At itel, designers are well aware that their users are the "bottom of the pyramid" population in "low-end markets." Here, designers consist of two types of people: one is product managers in charge of product planning, and the other is engineers who resolve software and hardware issues at the technical level. Generally speaking, there are two main ways for itel designers to know their customers. The first is through direct visits. For example, during his visit to Nigeria, product manager Y discovered why mobile-phone cost was still their customers' top consideration. There, he encountered a Nigerian farmer who saved money for almost one year to buy an entry-level itel smartphone (about US$50.00). The second way is through the translation of marketers and analysts. In general, itel marketers classify African markets into five tiers: tier one refers to capital cities, tier two is made up of small and medium-sized cities, and tiers three to five consist of a large number of towns and villages. In this way, itel users are translated as "people living in tier three to five markets," or simply, "the rural people." Except for geographic location, itel marketers further differentiate customers by their age, gender, occupation, and disposable income. As marketing manager L described, itel feature phone users are likely to be farmers, manual workers, and street vendors, while smartphone users are usually young students and office workers. To gain local knowledge, itel hires a small team of analysts to conduct market and user research. They use surveys, interviews, and participant observations to collect data. In both Accra and Kumasi, I met itel product analysts conducting interviews with promoters and users to ask their opinions about several itel prototypes. In another case, an itel user researcher stayed in a customer's home for a few days to investigate African users' "special needs," conducting what we may call "mini-ethnography." Although such an approach is not invented by itel, what is special about itel is its involvement of a significant number of local translators, especially marketers, in this process.

Through the labor of marketers and analysts, a large amount of local information can be collected, but only part of it will be selected and translated into the design network. In Kumasi, after quite a few promoters expressed their preference for a particular prototype, I asked the product analyst whether he would choose that one as the new model for 2020. "No," he says. "It is good but also more expensive. We may consider it in the future when our customers are able to pay more for such design." In this case, what matters for translators is not design per se, but design that matches customers' purchasing power during a certain period. At itel, it is common for a design idea to be rejected because of the cost. Product manager Z explains itel's product development process in this way:

> We start projects like this: a US$20 feature phone or a US$50 smartphone. Say we are going to design a US$50 smartphone for African markets, its main "specs" [specifications] immediately come into my mind: what kind of chipset, camera, or screen, how much memory, how big battery, etc. These specs are relatively stable

for smartphones at a certain price level. You must have these. Otherwise, your products will lose competitiveness.... These key components account for almost 90% of the phone cost. There is not much space left for us. We can only squeeze cost in some "small parts." If you want to add more features to attract customers, you have to delete others you think unimportant. You always need to make choices.

According to Z, the target cost will be set at the beginning of every project. Mobile phone specifications may vary from model to model, but as long as the price is set, its specifications are relatively stable due to market competition. The differences between low-end phones mainly lie in the "small parts"—small features that are attractive but not very expensive. It is usually these parts that are subject to negotiation. In Shenzhen, I once witnessed itel designers debating about whether to add a loudspeaker that cost only a few cents to a feature phone model, and whether to increase its phone battery from 1500 mAh to 1900 mAh. "For us, every cent counts," says one product manager. As it is not easy to strike a balance between maintaining costs and meeting demands, tensions may arise among various types of designers, or between designers and translators. For example, most itel software engineers have complained to me about the difficulty of adding all the features requested by product managers. As software engineer J says,

> Many itel smartphones have only 1GB of RAM [random-access memory]. Can you imagine that?! The [Android] operating system has already taken a lot of space. We also need to install some built-in apps. Sometimes product managers will ask us to add more fancy features, like face unlock or fingerprint. This makes our work very difficult. You only have 1GB. How can you put so many things (in it)? It will slow down the system. Users will hate that. So, we have to do a lot of optimization work.

For itel software engineers, much optimization work would be unnecessary if the phone had better hardware. Although itel hardware engineers also acknowledge this, they insist that their budget at the current stage can only afford "good enough" hardware. For example, phone memory is the most expensive part of many itel models. "Right now, its supply chain is monopolized by a few memory manufacturers like Samsung and Micron," an itel hardware engineer says. "When the phone price is pre-set, we have no other choice unless cheaper memory suppliers emerge." In this case, the internal tension among itel designers is mainly caused by an external force—monopoly suppliers—and thus, it can only be resolved when the supply chain has been diversified. In other cases, the pressure on itel engineers comes from product managers or marketers. For example, when designing itel 5606, itel engineers were asked to install both wireless radio and Facebook to this feature phone model. According to itel marketers, they added these two features because, for many itel users, radio is the most important means of entertainment and "Facebook is the internet." In other words, their translation of the market situation suggests that itel 5606 must have these two features in order to be attractive and competitive.

Based on the above analyses, this study further adjusts the analytical framework by adding another two important actors into the design network—suppliers and other designers (designers outside Transsion), both of which can shape how technologies are designed (Arrows 8 and 10, Figure 3). Due to market competition, Transsion designers and other designers can influence one another through their design choices (Arrow 7). As shown in the memory case, how designers can write ideas into technologies is also shaped by the availability of suppliers (Arrows 9 and 11).

In the itel case, on one hand, its design practice is structured by fierce market competition and a tight budget; on the other hand, it also benefits from Shenzhen's design ecosystem, including the rapid prototyping and "modularization" of mobile-phone production.[40] When designing phones, itel buys core technologies from other companies through "modules," integrates them with other components,

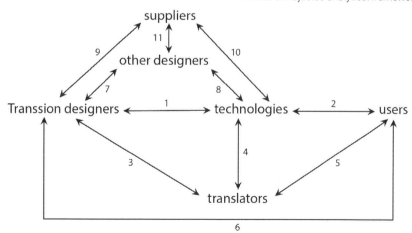

FIGURE 3. Adjusted analytical framework

and adds features to create new models. The low price of itel phones is also made possible by the emergence of Chinese and Taiwanese suppliers in the upstream of the supply chain. As the high patent fees of Qualcomm chipsets far exceed itel's budget, it resorts to MediaTek and Spreadtrum to develop technological solutions for low-end phones. In 2017, Google released its Android Go edition to tap into the low-end markets. As an operating system tailored for devices with less than 2GB of RAM, Android Go is widely used by itel phone models. These examples demonstrate that, apart from tension and competition, cooperation could also be a potential relationship between Transsion designers, other designers, and suppliers.

For African users, efforts made by various designers and translators have opened up new possibilities for their daily interactions with mobile technologies. With affordable feature phones, Kelly can use the flashlight to cook dinner, her mother can listen to local radio programs, and many other villagers can access the internet through Facebook. After information about local demands and contexts is translated into the design network and written into technologies, African users will adapt and appropriate these technologies in their everyday lives. According to my fieldwork, Ghanaian users have developed creative ways to deal with the various hardware and software limitations: they will reduce screen brightness to save battery, turn off the data connection when they are not using it, and store photos in extra memory cards instead of in their phones. These demonstrate that BOP users are not passive, but able to "read" and in some cases "revise" technologies in their social contexts (Arrow 2, Figure 3).

To summarize, itel's design practice has been shaped by various actors and factors, making its profit margins highly structured. In 2018, Transsion earned RMB¥22.5 billion by selling 124 million mobile phones, of which 72.6 percent (90 million) were feature phones; in contrast, Huawei earned RMB¥348.9 billion by selling 206 million smartphones in the same year.[41] The contrast is more revealing if we consider Apple: in the final quarter of 2017, when Apple accounted for only 19 percent of global smartphone shipments, it captured 87 percent of the profits.[42] Itel employees are acutely aware of these profit differences and their structural position in the global industry. As several interviewees pointed out, it is because itel lacks core technologies that it decides to focus on the relatively less attractive and profitable BOP markets. Historically speaking, this strategy is not new for Chinese ICT companies. When Ningbo Bird started as a township enterprise in the 1990s, it bypassed the saturated markets in big cities and adopted a low-price strategy to secure low-end markets, first in China's third- and fourth-tier cities and towns and later in overseas markets such as Africa. George Zhu, Transsion's cofounder and chief executive officer, was the former director of Ningbo Bird's overseas business unit. Before founding Transsion, Zhu traveled to ninety countries to promote Bird phones.[43] To some extent, what itel has been doing is translating this low-price strategy from China to Africa. What has made such translations possible is a clear understanding of itel's structural position in the global ICT industry and its customers' socioeconomic position in the global economy.

Dual/Multiple SIM Cards: Translating a *Shanzhai* Idea

The function of dual/multiple-SIM cards was often associated with *shanzhai* phones a decade ago. It was first developed for Hong Kong and Taiwanese entrepreneurs who traveled frequently to Shenzhen (Arrow 8, Figure 3).[44] Transsion appropriated this design idea in its expansion into Africa for various reasons (Arrow 7, Figure 3). During his interview with Chinese media, Transsion's vice president Arif Chowdhury explains that Africans need more than one SIM card because they will be charged more if they call via different telecom networks.[45] In Ghana, this is only one of the several reasons for the popularity of this function.

Ghana began privatizing its telecom sector in the 1990s and generally adopted a free market approach.[46] In 2019, the three largest telecom operators in Ghana were MTN, Vodafone, and Airetel-Tigo. Although it is not easy to figure out how the tariff system works in different scenarios, Ghanaian users are aware that it is more expensive to call across networks. For example, Vodafone has a two-*cedi* (US$0.34) daily package called "Red One," which offers 50 MB of internet data, unlimited calls to Vodafone numbers, but only 10 minute calls to other local networks. Because it only takes two *cedis* to get a new SIM card, Ghanaians tend to buy two or more cards to reduce the chance of calling across networks. Some users also claim that the voice quality is better when calling within the same network. For others, strong and reliable internet is an important factor to consider. When the internet from one network is poor or discontinuous, which is common in Ghana, users can switch to another network.

Moreover, due to market competition, telecom operators often provide diverse voice and data bundles and special offers to attract customers. With more SIM cards, users can switch to the bundle choice that best suits them at a particular moment. A young man from Kumasi told me that he often downloads movies during the night through MTN's "midnight bundle," which gives him 2.79GB data for one *cedi* and 5.33GB for three *cedis* from 12am to 5am. This phenomenon is not unique to Ghana. Pype's observation in Kinshasa also reveals a particular economy of "digital nightlife," in which most Kinois take shortcuts to online life through cheap midnight bundles.[47] In other cases, having more SIM cards makes it possible to avoid the pitfalls hidden in the complicated packages offered by telecom operators. An Uber driver in Accra claims that MTN internet works better in certain districts of Accra, while Vodafone internet works better in some other districts. His perception may not be entirely accurate, but from an infrastructural perspective, it indicates the uneven distribution of base stations in Accra, which is a widely acknowledged reality. He also insists that telecom operators will reduce the internet performance for weekly and monthly bundles because they are relatively cheaper. Therefore, he always prefers hourly and daily bundles. These cases suggest the diverse interpretations and appropriations of dual/multiple-SIM function among Ghanaian users in their everyday lives (Arrow 2, Figure 3).

In Transsion's efforts to bring dual/multiple-SIM phones to Ghana, there exist at least two layers of translation: one technical and the other symbolic. Technically, it is not difficult to translate this design idea from *shanzhai* phones to Transsion phones. However, doing this makes Transsion phones symbolically associated with the low-quality *shanzhai* phones, especially in the early stages. As a local phone retailer from Tema remarks,

> Many Ghanaians had the perception that an original phone should have a single SIM. The rationale behind it is that all original phones were single SIM until fake dual- and multiple-SIM phones were introduced. It was in most people's mind until some well-known brands also started to have this function. Now, even Apple has dual-SIM phones!

Another phone retailer in Kumasi says:

> People had doubts when we brought in itel and Tecno [phones]. They didn't trust "China phones" very much. I told my customers they had strong battery and other advantages. After they tried them, they found that they were different from other China phones.

Their words suggest that local retailers reshaped Ghanaians' perceptions of dual/multiple-SIM function by delinking it from fake "China phones" on one hand

and relinking it with other well-known brands like Apple on the other hand. In this process, the translation work of local retailers played an important role in reshaping the symbolic meanings of dual/multiple-SIM function (Arrows 4 and 5, Figure 3). In current Africa, dual/multiple-SIM function has become the norm rather than the exception. In 2018, around 87 percent of Kenyan users used multi-SIM smartphones, and the figure was around 70 percent in countries like Ghana, Nigeria, and Egypt.[48] My visits to mobile-phone markets in Ghana revealed that even Samsung and Apple brought dual-SIM support to some of their phone models. In 2018, Apple also released unique iPhone models with two nano-SIM cards for the Greater China Region to meet Chinese people's high demand for this function. This demonstrates that a design solution for a certain locality can challenge and reshape regional or even global design norms.

Battery: Translating Local Infrastructures

According to Ghanaian phone dealers, it was a strong battery that helped Transsion brands quickly gain popularity when they entered Ghana. My interviews with Ghanaian users also confirm that a strong battery is a significant reason to buy a Transsion phone. Even now, long-lasting battery remains a major selling point that distinguishes Transsion phones from other brands in the market. For example, most itel power-series smartphones have a 4000–5000 mAh battery. When itel launched its P32 smartphone model in 2018, the marketing slogan was "One Charge for Three Days." In comparison, iPhone XR and Samsung Galaxy J2, which were also released in 2018, have battery capacities of 2942 mAh and 2600 mAh, respectively. If this suggests that low-end users have higher demands for battery capacity than elite users, what is the reason for it? According to marketing manager W,

> Small battery is a "pain point" for many Africans. You won't understand this if you don't go there. The environment is totally different [from Shenzhen]. Blackouts are very common, even in capital cities. I encountered them quite a few times when I was in Nigeria. Many hotels and rich people have their own [electric] generators. But rural people are not so lucky. Many still don't have access to electricity. It's not that easy to charge their phones.

His description links African demand for big batteries to the local electricity situation. According to the International Energy Agency, sub-Saharan Africa had an electrification rate of 43 percent in 2017; due to uneven development, the electrification rate in rural areas was only 28 percent, much lower than that in urban areas (67 percent).[49] In the case of Ghana, its electricity sector is heavily dependent on hydropower generated from the Akosombo Dam built in 1965. Due to the uncertainty of rainfall and water inflows, Ghana suffered from serious electric power rationing in the years 1983–1984, 1997–1998, 2003, and 2006–2007.[50] The frequent blackouts have led local people to coin the term *dumsor*, meaning "off and on," to describe their everyday experience during the power crisis. In recent years, Ghana has increased its installed generation capacity by introducing an array of thermal power plants running on natural gas. The percentage of people with access to electricity in Ghana increased from only 15–20 percent in 1989 to 82.5 percent in 2016, which ranked high in sub-Saharan Africa.[51] Having a house connected to the national grid, however, does not guarantee a reliable and affordable supply of electricity. In 2019, I experienced blackouts in both Accra and Kumasi, which were caused by an unreliable supply of natural gas. My visit to Kelly's village in Kpone showed that the urban-rural gaps persist, and Ghana has a long way to go to achieve universal and reliable access to electricity.

At itel, many people are aware of the widespread power shortage in sub-Saharan Africa because they (including some designers) have made short- or long-term visits there (Arrow 6, Figure 3). In many other cases, it is translators like marketing managers who bring back information about local infrastructures (Arrows 3 and 5, Figure 3). How will such information be translated into the phone design? As hardware engineer S explains:

> When we design phones, we see both visible and invisible things. What's the visible thing? Taking the battery for example, how big and how heavy [is it]? Our customers need a big battery, but it will occupy more [physical] space. To save space for it, we need to make sacrifices in other parts, and the overall performance will be influenced. The phone will become heavy and ugly. That's probably why iPhone designers don't like big batteries. . . . Then, what's the invisible thing? Protocols, standards, norms, etc. For example, to ensure a phone can get signals, the power level can range from 30 dB to 33 dB. In places like Shenzhen, 30 is enough because almost every corner is covered by mobile phone base stations, but in Africa, I will choose 33. Because there are not enough base stations there, you need a higher power level to get signals.

His narrative indicates that itel's design practice has been shaped by customer demands (e.g., a big battery), material properties (e.g., the size and weight of the battery), industrial standards (e.g., signal strength), and local infrastructure (e.g., base station). If this suggests that translating local knowledge into the phone design is not a straightforward process but full of negotiations, how can we know if local knowledge has been "successfully" written into technologies? In other words, how can we know if a translation is a "good" translation? One possible way to answer this is by seeing through the eyes of users. In Kintampo, there is a shop owner who has been selling itel phones for seven years. Showing me the popular itel 5606 (see Figure 2), he explains why local people like this brand:

> Their products are good. The battery is good. The network is good, especially this one [itel 5606]. When they [his customers] go to villages, the network is good. So, they all come to buy this one. For some other phones, no network. This one, I can sell 20 pieces a day.

If we view itel 5606 as an example of a good translation, it is because engineers have made tradeoffs elsewhere to save space for its battery, and modified "invisible things" like power levels to ensure that rural users can also get signal reception despite the lack of base stations in their villages. The design experiences at itel demonstrate that the actual, concrete form of artifacts does not preexist in the maker's mind but "grows" from the mutual involvement of people and materials and from the gradual unfolding of local and global forces.[52] By relying on various translators or designers' own visits, itel makes it possible to integrate local knowledge into the design process, which often involves modifications to and interventions in existing standards and norms. Such an approach represents a meaningful departure from the "digital universalism" of Western design,[53] especially humanitarian design, in which technologies designed in the labs of digital centers are expected to solve the problems in the peripheries.

Camera: Translating the Beauty of Darker Skins

In Africa, Transsion has widely promoted its optimized camera for darker skin tones as a key selling point. It developed this strategy because many existing camera phones could not provide Africans with clear, high-quality selfies. The difficulty of photographing darker skin tones is not a recent problem, but has long been embedded in various visual technologies.

In the era of still photography, Kodak's Shirley cards, the norm reference cards using "Caucasian" female models, became the industry standard for calibrating skin tones in the 1940s and 1950s.[54] As Roth points out, the look and light skin tones of these models conformed to a popular masculinist notion of beauty, which was largely defined from a Western male perspective.[55] In the era of color film and television, the research agenda and the choice of chemicals were also dominated by the need to reproduce Caucasian skin tones.[56] Given the assumption that the rightful subjects would be white, Western designers inscribed a light-skin bias into visual technologies from the beginning. The result is that when images of non-Caucasian skin tones are projected in visual media, they are often distorted or highly deficient. In 1978, the French filmmaker Jean-Luc Godard refused to use Kodak film to shoot in Mozambique because the light range was so narrow that "if you exposed film for a white kid, the black kid sitting next to him would be rendered invisible except for the whites of his eyes and teeth."[57] When Kodak began to incorporate more diverse models in the 1970s, it was not due to the pressure from the black community but because its two biggest clients—the confectionary and furniture industries—complained that the subtle differences of dark chocolate and dark furniture could not be discerned.[58]

Today, visualization is a way of "computer-aided seeing,"[59] which is increasingly mediated by computational algorithms. A growing body of literature has documented the racial bias of face recognition algorithms, from Hewlett-Packard webcam's failure to detect African American faces to Google's auto-tagging pictures of black people as "gorillas."[60] In 2016, "Beauty.AI," the first international beauty contest decided by an algorithm, sparked controversy after the results revealed that the robots did not like people with dark skin.[61] This is mainly because the datasets used to train the algorithms are not diverse enough, and minority groups are often underrepresented.

In these cases, although designers are not always conscious, they have practiced discriminatory design in which a light-skin bias is written into the deep structure of visual technologies (Arrow 8, Figure 3). To undo these biases, Transsion designers need to "revise" or "rewrite" the fundamental parts of these technologies—the dataset and the algorithm (Arrow 1, Figure 3). To combat the underrepresentation of dark-skinned people in previous datasets, Transsion collected millions of images of dark-skinned people, created its own machine learning dataset, and used it to train its algorithms. Transsion established a research team at the company level and collaborated with various partners, including ArcSoft, SenseTime, and Visidon, to develop camera technologies that can capture darker skin with accuracy and sensitivity.[62] When explaining their camera design practice, an algorithm engineer at Transsion says,

> The problem we need to tackle is not just on the technical level. Instead, what we need to do is to understand the beauty of dark-skinned people. What Africans want is not whiteness but beauty. What they want is a better self.

This indicates that designing cameras for dark-skinned people requires not only computing knowledge but also understandings of beauty in different social-cultural contexts. In other words, for Transsion designers, "technologies of seeing" and "ways of seeing" are not separate, but closely related processes. In another case, when introducing itel S15 on its official website,[63] itel presents several images of young, dark-skinned female models, saying:

> The AI algorithm identifies your facial features, skin tone and lighting environment to add beauty effects tailored to individual facial features and bring out your unique natural beauty.

But what is "natural" beauty? How can designers ensure that they have translated the correct criteria of beauty into technologies? If natural beauty is the goal, do such criteria exist at all? As these camera algorithms are often proprietary, it is very difficult to know how engineers have made specific decisions at various stages. However, my interviews with various designers and marketers suggest that Transsion employees do not have a unified understanding of "the beauty of dark-skinned people," although many of them refer to it. In describing how to "beautifully" present African faces, some interviewees say it should be "bright" and "clear," while others say it should be "light brown." One of the advertisements for itel S13 claims its "Face Beauty" camera can make the face look "slim," "smooth," and "ruddy," giving you "stunning chocolate skin." That the face should be "slim" and "smooth" indicates that, in some areas, Transsion is conforming to, instead of challenging, mainstream beauty norms, especially those in the East Asian context.

That said, Transsion designers and translators have attempted to embrace the diversity of skin colors and present their subtle differences through their design and translation practices. This could be viewed as a celebration of multiculturalism in a commercial context. In Shenzhen, itel marketers often emphasize that Africa is not monolithic but consists of fifty-four countries with different cultures, tribes, and races. Therefore, it is almost impossible to translate a universal standard of beauty. At itel, software engineers frequently use the term "optimization" to describe their work. From an etymological perspective, the term "optimize" derives from the Latin *optimus* ("best"); in computing, to optimize is "to change data, software, etc. in order to make it work more efficiently or to make it suitable for a particular purpose."[64] In this sense, it will be meaningless to talk about the "best" solution before the particular purposes or target users are identified. For example, I used itel S13 for a few months during my fieldwork. For Asian users like me, its camera is very clear but somewhat too bright.

Regarding Ghanaian users, although very few of them are aware of the camera differences at the algorithmic level, it does not mean they cannot "read" these differences at other levels. For example, after receiving an iPhone from her boyfriend as a gift, a young promoter in Accra still uses her Tecno phone to take selfies. "Somehow, the Tecno camera works better for me than my iPhone," she says. "It is clear. It looks like me." Just like her, many Ghanaians mentioned "clarity" as an important criterion for a good camera. What "clarity" means may vary from person to person, but it points to Ghanaian users' frustrating experiences with other cameras that have failed to photograph black skin tones with clarity. In this sense, Transsion's camera technologies open up new possibilities for dark-skinned users to capture and embrace their own aesthetics of beauty through less discriminatory, if not totally fair, algorithms, which has been made possible through a series of translation processes.

Conclusion: Toward a Southern Perspective of Design

Focusing on the case of itel, this chapter examines how it designs affordable phones for Africa's BOP population. Using "technology translation" as an analytical framework, the study illustrates itel's design network, which is comprised of diverse actors including designers, translators, upstream suppliers, and downstream users. Through the cases of SIM cards, battery, and camera, this chapter elaborates on the dynamics and tensions of translating and designing low-cost phones for African users. Shaped by various factors like market competition, global supply chain, local culture, and infrastructure, these processes often involve challenging, modifying, and revising existing design standards and norms.

Being cost-sensitive, context-conscious, and demand-driven, itel develops a design approach and business model based on both its "place" in the global ICT industry and its customers' "place" in global consumer society. In so doing, it creates a disruptive force that transforms the previously "unusable" BOP market into a mass market. Adopting a historical perspective, this study argues that the business model of reducing profits to gain market share is not unique to Transsion but familiar to many Chinese ICT companies. This has been shaped by two main factors. The first is the informational stratification and market segmentation that pervades the Global South. Due to distribution barriers and low profit margins, Third World rural markets have been largely ignored by large corporations. The second factor is the structural inequality of the global ICT industry. Monopolizing core technologies, Western high-tech companies tend to pursue an urban-centric, elite-oriented, and high-value-added business model. To avoid competition, companies from peripheral countries have little choice but to start with low-end markets. Even Huawei had a humble beginning of targeting China's rural markets, adopting a strategy called "encircling cities from the countryside."[65] What is unique about the Transsion case is how it translates such an approach into the African context, which involves a large number of translators offering their creative labor and local knowledge.

As scholars have noted, many current design practices tend to reproduce the "one-world" project of neoliberal globalization, which not only reduces differences but also eliminates people's abilities to imagine being otherwise.[66] According to Tony Fry, modern design as a "defuturing" project has played a critical role in the expansion of Western modernity and its systematic creation of unsustainability.[67] For Mignolo, design is an imperial project in which the colonial matrix of power has become the global design of Western civilization.[68] For him, "otherwise" is equivalent to decolonial thinking and doing. To recover our future-imagining capacity, we must first unlearn the defuturing traps of modern design; "learning to unlearn is crucial to guide us in re-learning and to be/do otherwise."[69] In this process, a Southern perspective of design is an important step in bringing in Southern knowledge and nourishing alternative socio-technical worlds.

As a counter-example of Western design, itel provides an important case to rethink the geopolitics of design. Deeply rooted in Shenzhen's maker culture, itel's design strategy is both a response to and an adjustment of the structural inequality of the global ICT industry. It demonstrates that Southern countries are neither passive nor monolithic, but able to assert different visions of digital futures through design and innovation, although such futures are often uncertain due to the paradox of disruption/structure. However, it would be inaccurate to say itel is *the* alternative we are looking for. After all, it belongs to a private company aimed at making a profit. That itel will introduce about fifty models every year to boost sales suggests that it has not escaped the "defuturing" traps. When its parent company, Transsion, challenges the cultural imaginary of Silicon Valley as the center of future-making for everywhere, it is also making Shenzhen the next center of future-making for Africa. It is possible for Shenzhen to be Silicon Valley's "South" and Africa's "North" at the same time. In this sense, the South is not only a plural but also a relative concept. To deploy the critical lens of integration/differentiation, Transsion as a case problematizes the North-South binary and demonstrates the complexity and "multipolarity" of communication innovation. From Kenya's iHub to Nigeria's Co-Creation Hub, innovation incubators are springing up across the African continent. Will we have African designers to challenge Shenzhen and/or Silicon Valley? Can we design plurally and think otherwise? To achieve this, it is important to engage with various "Souths" and their distinctive digital futures.

NOTES

This chapter is a partial reprint of "Designed for the Bottom of the Pyramid: A Case Study of a Chinese Phone Brand in Africa," Miao Lu, *Chinese Journal of Communication*, April 2020, https://doi.org/10.1080/17544750.2020.1752270. Copyright © The Communication Research Centre, The Chinese University of Hong Kong, reprinted by permission of Taylor & Francis Ltd, http://www.tandfonline.com on behalf of The Communication Research Centre, The Chinese University of Hong Kong.

1. See Transsion's annual report for 2019: https://mp.weixin.qq.com/s/4XwgM7GKCUe6a9-AOXQMfw/.
2. James Ferguson, "Seeing Like an Oil Company: Space, Security, and Global Capital in Neoliberal Africa," *American Anthropologist* 107, no. 3 (2005): 377–82.
3. Brian Larkin, *Signal and Noise: Media, Infrastructure, and Urban Culture in Nigeria* (Durham, NC: Duke University Press, 2008); Ravi Sundaram, "Recycling Modernity: Pirate Electronic Cultures in India," *Third Text* 13, no. 47 (1999): 59–65.
4. Anita Chan, *Networking Peripheries: Technological Futures and the Myth of Digital Universalism* (Cambridge, MA: MIT Press, 2013).

5. Coimbatore K. Prahalad, *The Fortune at the Bottom of the Pyramid: Eradicating Poverty through Profits* (Upper Saddle River, NJ: Wharton School Publishing, 2005).
6. Michael Blowfield and Catherine S. Dolan, "Business as a Development Agent: Evidence of Possibility and Improbability," *Third World Quarterly* 35, no. 1 (2014): 22–42.
7. Walter D. Mignolo, "Epistemic Disobedience, Independent Thought and Decolonial Freedom," *Theory, Culture & Society* 26, no. 7–8 (2009): 159–81.
8. Seyram Avle and Silvia Lindtner, "Design(ing) 'Here' and 'There': Tech Entrepreneurs, Global Markets, and Reflexivity in Design Processes," in *Proceedings of the 2016 CHI Conference on Human Factors in Computing Systems* (Denver: ACM Press, 2016), 2233–45; Lilly Irani, "'Design Thinking': Defending Silicon Valley at the Apex of Global Labor Hierarchies," *Catalyst: Feminism, Theory, Technoscience* 4, no. 1 (2018): 1–19.
9. Arturo Escobar, *Designs for the Pluriverse: Radical Interdependence, Autonomy, and the Making of Worlds* (Durham, NC: Duke University Press, 2018), 205–7.
10. Escobar, *Designs for the Pluriverse*, xvi.
11. See "Transsion, the King of Mobile Phone in Africa," https://pandaily.com/transsion-king-mobile-phone-africa-makes-fortune-low-profile/.
12. See Xinhua Headlines: "China's Call for Globalization," http://www.xinhuanet.com/english/2020-01/23/c_138729714.htm/.
13. For "Going-out" policy, see http://www.chinadaily.com.cn/a/201901/02/WS5c2bfd17a310d91214051f36.html. Also see Jia and Nieborg in this volume for a discussion of the Digital Silk Road.
14. Barry M. Katz, *Make it New: A History of Silicon Valley Design* (Cambridge, MA: MIT Press, 2015).
15. Katz, *Make it New*, 70.
16. Irani, "Design Thinking"; Katz, *Make it New*; Lucy Kimbell, "Rethinking Design Thinking: Part I," *Design and Culture* 3, no. 3 (2011): 285–306.
17. Bruce Mau et al., *Massive Change* (London: Phaidon Press, 2004), 11.
18. Kimbell, "Rethinking Design Thinking."
19. Irani, "Design Thinking."
20. Jack Linchuan Qiu, *Goodbye iSlave: A Manifesto for Digital Abolition* (Champaign: University of Illinois Press, 2017).
21. Kelly Hu, "Made in China: The Cultural Logic of OEMs and the Manufacture of Low-Cost Technology," *Inter-Asia Cultural Studies* 9, no. 1 (2008): 27–46.
22. Lindtner et al., "Designed in Shenzhen: Shanzhai Manufacturing and Maker Entrepreneurs," in *Proceedings of the Fifth Decennial Aarhus Conference on Critical*

Alternatives (Aarhus, Denmark: Aarhus University Press, 2015), 85–96; Cara Wallis and Jack Linchuan Qiu, "Shanzhaiji and the Transformation of the Local Mediascape in Shenzhen," in *Mapping Media in China*, ed. Wanning Sun and Jenny Chio (London: Routledge, 2012), 109–25.

23. Lindtner et al., "Designed in Shenzhen."
24. Yuqing Xing, "Global Value Chains and the Innovation of the Chinese Mobile Phone Industry," *East Asian Policy* 12, no. 1 (2020): 95–109.
25. See Mutibwa and Xia in this volume.
26. Jack Linchuan Qiu, *Working-Class Network Society* (Cambridge, MA: MIT Press, 2009).
27. Chuan-Kai Lee and Shih-Chang Hung, "Institutional Entrepreneurship in the Informal Economy: China's Shanzhai Mobile Phones," *Strategic Entrepreneurship Journal* 8, no. 1 (2014): 16–36.
28. Lindtner et al., "Designed in Shenzhen."
29. See "China's Innovation Drive," http://www.chinadaily.com.cn/china/2011-03/11/content_12156650.htm/.
30. Xudong Gao, "A Latecomer's Strategy to Promote a Technology Standard: The Case of Datang and TD-SCDMA," *Research Policy* 43, no. 3 (2014): 597–607.
31. Gao, "A Latecomer's Strategy to Promote a Technology Standard."
32. Yuezhi Zhao, "China's Pursuits of Indigenous Innovations in Information Technology Developments: Hopes, Follies and Uncertainties," *Chinese Journal of Communication* 3, no. 3 (2010): 266–89. See Lianrui Jia and David Nieborg in this volume.
33. Lee and Hung, "Institutional Entrepreneurship."
34. See "Transsion's Lead in African Phone Market under Threat," https://technode.com/2019/04/22/transsions-lead-in-african-phone-market-under-threat-from-fellow-chinese-rivals/.
35. Xing, "Global Value Chains."
36. Steve Woolgar, "The Turn to Technology in Social Studies of Science," *Science, Technology & Human Values* 16, no. 1 (1991): 20–50; Madeleine Akrich, "The Description of Technical Objects," in *Shaping Technology/Building Society: Studies in Sociotechnical Change*, ed. Wiebe E. Bijker and John Law (Cambridge, MA: MIT Press, 1992), 205–24.
37. Jenna Burrell, *Invisible Users: Youth in the Internet Cafés of Urban Ghana* (Cambridge, MA: MIT Press, 2012).
38. During these three months, I worked as an intern at Transsion's itel department. I informed itel of my future research intention and received no salary from it.
39. In 2019, the average price of smartphones worldwide was US$302. See https://www.

statista.com/statistics/788557/global-average-selling-price-smartphones/.
40. Xing, "Global Value Chains."
41. See "Transsion's Lead in African Phone Market under Threat."
42. See "Apple Continues to Dominate the Smartphone Profit Pool," https://www.forbes.com/sites/chuckjones/2018/03/02/apple-continues-to-dominate-the-smartphone-profit-pool/#47bdfcae61bb.
43. See "Transsion's Lead in African Phone Market under Threat."
44. Wallis and Qiu, "Shanzhaiji and the Transformation of the Local Mediascape in Shenzhen."
45. See "Transsion, the King of Mobile Phone in Africa."
46. Luke Haggarty et al., "Telecommunication Reform in Ghana," *World Bank Policy Research Working Paper* 2983 (2002).
47. Katrien Pype, "(Not) in Sync: Digital Time and Forms of (Dis-)Connecting: Ethnographic Notes from Kinshasa (DR Congo)," *Media, Culture & Society*, Special Issue: Media and Time (2019): 1–16.
48. See "Transsion's Lead in African Phone Market under Threat."
49. See "Covid 19 Reverses Electricity Access Progress," https://www.iea.org/sdg/electricity/.
50. Ebenezer N. Kumi, *The Electricity Situation in Ghana: Challenges and Opportunities* (Washington, DC: Center for Global Development, 2017).
51. Kumi, *The Electricity Situation in Ghana*.
52. Tim Ingold, *The Perception of the Environment: Essays on Livelihood, Dwelling and Skill* (London: Routledge, 2000), 345.
53. Chan, "Networking Peripheries."
54. Lorna Roth, "Looking at Shirley, the Ultimate Norm: Colour Balance, Image Technologies, and Cognitive Equity," *Canadian Journal of Communication* 34, no. 1 (2009): 111–36.
55. Roth, "Looking at Shirley, the Ultimate Norm."
56. Brian Winston, *Technologies of Seeing: Photography, Cinematography and Television* (British Film Institute, 1996).
57. See "Racism of Early Colour Photography," https://www.theguardian.com/artanddesign/2013/jan/25/racism-colour-photography-exhibition.
58. Roth, "Looking at Shirley, the Ultimate Norm."
59. Min Chen and Luciano Floridi, "An Analysis of Information Visualisation," *Synthese* 190, no. 16 (2013): 3421–38.
60. Ruha Benjamin, *Race after Technology* (Cambridge: Polity Press, 2019).

61. See "A Beauty Contest Was Judged by AI," https://www.theguardian.com/technology/2016/sep/08/artificial-intelligence-beauty-contest-doesnt-like-black-people.
62. See "Transsion Holdings' Journey to an IPO," https://www.counterpointresearch.com/transsion-holdings-journey-ipo-strengths-challenges/.
63. See "16MP Brighter Selfie," https://www.itel-mobile.com/global/products/smart-phone/s-selfie/s15/.
64. See Oxford Learners' Dictionaries, https://www.oxfordlearnersdictionaries.com/.
65. Yun Wen, *The Huawei Model: The Rise of China's Technology Giant* (Urbana: University of Illinois Press, 2020).
66. Escobar, *Designs for the Pluriverse*; Tony Fry, *A New Design Philosophy: An Introduction to Defuturing* (Sydney: UNSW Press, 1999).
67. Fry, *A New Design Philosophy*.
68. Walter Mignolo, *Local Histories/Global Designs: Coloniality, Subaltern Knowledges, and Border Thinking* (Princeton, NJ: Princeton University Press, 2012).
69. Eleni Kalantidou and Tony Fry, eds., *Design in the Borderlands* (London: Routledge, 2014), 182.

The Necropolitics of Innovation

Sensing Death in the Mediterranean Sea

Monika Halkort

The combined impact of climate change, loss of biodiversity, industrial waste, and noise pollution have established the world's oceans as critical platforms for anticipating risks of premature deaths and extinction. Oceans cover 70 percent of the Earth's surface, are the planet's largest biosphere, and play a critical role in regulating the world climate.[1] Monitoring their "health" has become indispensable in the struggle against global warming. It has led scientists to put a wide range of instrument platforms into the high seas to observe rising sea levels, oceanic temperature, salinity, and ocean currents in support of climate research.[2] The information generated by these mobile platforms is supplemented by coastal radars, webcams, and Earth Observation satellites that collect comparative data in next to real time.[3]

The expansive network of environmental sensors and Earth Observation satellites provides a powerful example of how technical innovations can foster positive change in the face of looming threats and destruction. Without the use of these technologies, it will not be possible to secure a more sustainable, equitable, and inclusive future that responds to the human-induced risks, vulnerabilities, and harms to the Earth system up to this point. That said, the same environmental platforms are also used for border security and military operations that skillfully exploit the infrastructures of ocean monitoring for political surveillance and control.

The policy directive for the European border management system EUROSUR has just recently been updated to strengthen the interoperability of military, environmental, and Earth observation satellites for the dual purpose of border security and policing.[4] The directive lays out a comprehensive strategy for automating the information exchange between fishing and environmental agencies, the EU Earth Observation Program, Copernicus, and the main coastguard and border-security agency, Frontex. The overall goal here is to integrate the joint output of data collected by each agency into a centralized platform for risk analyses and situational reports aimed at strengthening Europe's border protection capabilities.[5] EUROSUR combines automated vessel tracking and detection capabilities, software functionalities for algorithmic anomaly detection, with precise weather and oceanographic forecasts. Integrated into an automated information-exchange system, these functions significantly enhance Frontex's ability to locate and intercept vessels suspected of people or weapons smuggling and to share risk scenarios across EU member states. A similar convergence of military and scientific agendas can be witnessed in Chinese and American research activities in and around the Indo-Pacific. Both are critically dependent on oceanographic data drawn from image satellites, floating instrument platforms, and underwater equipment to support their overt and covert reconnaissance missions, in search of oil and gas but also in gathering intelligence about foreign submarines or secret military installations in the South China Sea.[6]

The push towards greater system integration of military and environmental technologies for advancing geopolitical and strategic interests is indicative of the persistent hold of military and defense logics over the production of scientific knowledge, in particular with regard to ocean research. Whereas it was once ships that functioned as main carriers of weapons and scientific instruments, it is today the vast network of image satellites, instrument platforms, and underwater drones directly communicating with orbiting radars that bring scientific knowledge and warfare into inextricable embrace. Their entanglement bespeaks a scientificization and weaponization of the environment and the Earth that powerfully manifests what Gabrys has called the "becoming environmental of computation."[7] Environments are made operational through sensors and instrument platforms, reconfiguring nature, the Earth, and the oceans into programmable entities. In what follows I explore how this programmability affects the ways risks of premature deaths at sea become knowable, sensible, and apparent, shaping possible responses by regimes of environmental protection and humanitarian care.

My discussion centers on the historically specific context of the Mediterranean, where risks of premature death have become particularly potent and differentially marked. The North Adriatic Sea has been identified as one of four hundred dead zones considered to be under acute threat from CO_2 emissions, industrial pollution, and human-induced changes in the marine ecosystem.[8] These environmental threats are overshadowed by a profound humanitarian crisis that has cost more than twenty-thousand migrant lives since 2014.[9] This figure includes the number of bodies known to have drowned, as well as those who are considered missing. No one can say for sure *if* they have died, or how they have disappeared, because data on migrant deaths are not systematically collected, quite unlike the amount of data gathered on dying species and marine habitats. And while not all migrant deaths are directly related to climate change and environmental destruction, they are undeniably linked to the ongoing series of wars and conflicts over the control of vital earth resources—i.e., water, fossil fuels, rare minerals, and deep-sea resources—that are key drivers of global warming, mass extinction, and human-induced environmental threats.[10]

The critical blind spot in the knowledge about migrant deaths testifies to the markedly uneven ways in which technological innovations are currently mobilized in response to vulnerabilities and harms shared across species. As I have argued elsewhere, they bespeak racialized habits of sense that deeply implicate computational infrastructures in biopolitical projects, where certain deaths no longer register and are placed outside the order of legal and moral responsibility and protection, and ultimately outside the spectrum of humanity as such.[11] Building on this previous work, this chapter interrogates the figural tactics of racialization in relation to the critique of innovation in a multipolar world.[12] My main aim will be to show how the ever more pervasive incursion of sensing technologies into the marine ecosystem recalibrates the fluid boundary between natural and premature death in techno-natural assemblages, while the distinction between environmental monitoring, intelligent warfare, and surveillance in these assemblages has become ever more difficult to ascertain and blurred.

What this chapter contributes to the general theme laid out in the introduction to this volume, then, is a more expansive reading of the idea of the "multipolar" from a post-humanist and non-anthropocentric perspective to draw attention to the multiplicity of nonhuman actors implicated in change and transformation and their effects across socio-technical and natural domains. Furthermore, by situating this debate in the Mediterranean rather than in Asia or the United States, this chapter also aims to open up a historical perspective to make room for acknowledging the slow violence of technical incursions premised on colonial myths of human ingenuity and progress through technical innovation under the pretext of universal betterment for all of humanity. This historical perspective serves as a premonition of ongoing and future projects of globalization and colonization of habitats that appeal to such myths.

I will start by situating the current infrastructure of marine science and Earth observation technologies in their colonial context and recall critical moments in the entanglement of military and scientific agendas with imperial regimes. The construction of the Suez Canal (1859) marks a critical turning point in this regard. It not only solidified the intimate link between military and scientific interests and desires, but also opened the sea to multiple processes of colonization and enclosure across Europe, Africa, and Asia, spreading the deadly logic of competition and conquest deep into multispecies life. Against this backdrop the canal stands in as powerful manifestation of the grand geo-engineering designs whose slow violence forcefully returns in the current climate crisis. The second part of the paper will show how the toxic afterlife of these techno-colonial incursions transformed some of the core ontological assumptions on which the biopolitical governance of life and death traditionally relied. It is here that the thousands of unidentified migrant bodies reveal deep inconsistencies in the ways new technologies for seeing, knowing, and engaging with the environment are utilized for containing the destructive effects of colonial capitalism. Rather than fostering equal recognition of the vulnerabilities and harms shared across species, I suggest, environmental sensing technologies ushered in a "post-human governance" of life and death that flexibly (re)distributes logics of racialization and dehumanization within and across species. By consequence, some deaths are naturalized while nature is historicized into an active agent in the anticipation of threats that declassifies certain bodies as improper to the terrestrial surface, an empty variable, that are left nondescript.

The empirical material for my argument was gathered through interviews with marine biologists, humanitarian and environmental activists, international NGOs, and representatives of NATO and the European border-security agency Frontex at the biannual Shared Awareness and De-confliction (SHADE) conference in 2017. These interviews are complemented with a close reading of mission reports, press releases, public media campaigns, scientific studies, and policy documents regarding environmental protection and border security in the Mediterranean Sea.

Rethinking Multipolar Innovation on a More-Than-Human Scale

Contemporary infrastructures of environmental technologies and Earth observation systems are deeply imbued with the exigencies of colonial capitalism and its military-scientific institutions. Together they provided the technological knowledge, expertise, and the high-risk capital necessary for investing in scientific and technical innovations on which colonial knowledge and domination were built.[13] Starting with Napoleon's Expeditions to Egypt (1798–1801), new measuring instruments, communication networks, and scientific methods (chronometer, telegraph, observatories) gradually transformed the Mediterranean into a site of ongoing scientific observation, "a living laboratory,"[14] where critical information about extreme weather patterns, disease, crop productivity, and desertification could be generated and mobilized for grand agricultural and climate engineering designs.[15] These experiments were decisive for the emerging geographies of European imperialism. It enabled those with the greatest technological advantage and military might to reallocate vital Earth resources and to construct a global world space in which previously separate biotas and raw materials could be flexibly mixed, circulated, and reassembled into new geological formations and natural habitats.[16] As Yusoff observes, "the afterlives of these geomorphic acts constitute the materiality of the Anthropocene and its natal moment"—from the transformation of the mineralogy of the earth through the extraction of gold, silver, salt, and copper to the massive transformation of ecologies as a result of transplanting people, animals, and plants.[17]

The construction of the Suez Canal (1859–1869) marks a pivotal moment in this succession of techno-natural incursions. It transformed the Mediterranean from a closed lake into a central corridor between Asia, Africa, and Europe, paving the way for ever larger parts of non-European lands to come under European control.[18] At the same time, it opened the sea to traveling species from Asia and the Indo-Pacific region, ushering in the biggest migration of marine life to date.[19] Connecting the Mediterranean with the Red Sea had long been a utopian dream, but also a tremendous challenge, because European engineers assumed that there is a 10 meter difference between the two seas.[20] It took several failed attempts before the project could be brought to fruition, further fueling its symbolic resonance as a token of Western ingenuity and technological mastery. As Valeska Huber recalls, the canal was envisioned as a crucial segment in the expansion of European civilization by way of modern transport and communication.[21] Three hundred years later, this universalizing promise manifests itself in rising sea levels, temperature increases, toxic algae concentrations, and oil spills that are threatening the survival of entire coastlines, cities, and the marine ecosystem from the Mediterranean to the South China Sea.

More than half of the one thousand known "non-indigenous species" (NIS) currently inhabiting the aquatic space connecting Libya, Italy, Greece, Turkey, and the Levant have entered via Suez, putting severe pressures on the genetic function and structure of local habitats.[22] They often multiply with extraordinary speed and overconsume or outcompete indigenous forms for space, food, and other vital resources.[23] Rising sea temperatures and global warming further accelerate these pressures, because tropical forms tend to thrive in warm waters, whereas local specimens and plants, which are already exhausted by industrial pollution and rising temperatures, are further weakened. Added to that comes the rapid increase in shipping traffic following the extension of the canal in 2014. Around 17,000 ships pass through Suez each year, moving tons of ballast water full of biological materials—including plants, animals, viruses, and bacteria—from place to place. Globally it is estimated that some ten thousand species circulate across the world this way at any moment,[24] changing oxygen levels, food chains, and other life-sustaining processes in local marine habitats.[25] In some instances, it was science itself that contributed to the spread of notorious biota. The "killer algae" *Caulerpa taxifolia*, for example, was initially imported to the Institute of Oceanography in Monaco for study purposes, but it quickly escaped into the Ligurian Sea through discharge pipes from the circulating seawater system, brushing aside relationships formed among Mediterranean plants and animals over tens of thousands of years.[26]

The precarious condition of the Mediterranean gives vivid testimony of the destructive impact of techno-scientific experiments and innovations on the "geo-logic" of the Earth system.[27] It created a political ecology of trans-species relations in which the toxic agential powers of waste, climate effects, and pollutants fold into evolutionary principles to become an integral feature of geological strata, climates, the Earth atmosphere, and biomass. Following Yusoff, I am using "geologic" here to denote how humans not only affect geology but are an intemperate force within it.[28] As Yusoff notes, "As geological agents, humans are explicitly located alongside other Earth and extraterrestrial forces that possess the power of extinction and planetary effect as a direct result of their ability to capitalize on and incorporate geological forces, making previous fossilizations such as oil, minerals or coal their own."[29] This immersion of human activity into geological strata and metabolic cycles fundamentally challenges some of the core ontological assumptions on which modern biopolitics traditionally relied.[30] It extends the governance of life and death into an environmental condition with the result that received boundaries and distinctions between nature and society, life and nonlife, bios and geos, which have been characteristic for modern-colonial government, lose their ontological grounds. As Lehman remarks, the question of securing certain forms of life can no longer be confined to the figure of the human, but rather "needs to respond to planetary-scale environmental conditions, not as a predictable baseline but as a source of potentiality that cannot be easily mapped or known."[31] To put it another way, as the folded temporalities of industrial pollution, aerosol, and CO_2 dispersion intermingle with biological and geological holds, long-standing assumption of what constitutes life, and the forces and conditions that sustain it become ever more difficult to ascertain, just as the time frames and the conditions of its ending are becoming ever more contested and unclear.

Against this horizon of uncertainty about risks and their "knowability," new logics of governance emerge that draw together diverse sets of knowledge and technologies that have never been in the purview of the biopolitical calculation, i.e., oceanography, atmospheric chemistry, Earth system science, or hydrography.[32] Combined with the speculative calculus of machine learning and big data analytics they conjure post-human ontologies of governing life and death within and across geo/biological strata, where previously fixed indicators for measuring health, security, or mortality are opened to recombinant correlations of variables stretched across vastly different time frames and scales.[33] Health today is as much a function of aerosol dispersion, rising sea temperatures, CO_2 emissions as it is a measure of access to vaccination and physical health care. These environmental qualities, Anderson suggests, demand a calculus of life and death quite different from the classical biopolitical imagination. Such a calculus can no longer rely on fixed variables known in advance, but rather it needs to account for unpredictable and emergent phenomena that are "potentially catastrophic and capable of altering the conditions of possibility for life across scales."[34]

Such a reading of governance as heterogenous assemblage of human and nonhuman actors and agencies calls for a more expansive notion of "multipolar innovation"—one that includes the multiplicity of substances, energies, objects, and materials implicated in the disruptive impact of technological inventions and scientific novelties. Satellite imaging and sensor technologies have introduced a whole new range of methods and means for observing metabolic life cycles through the medium of electromagnetic waves that fundamentally change how the fluid boundaries and transitions between life, death, and nonlife are interrogated, measured, and understood. Thanks to machine sensing and vision, scientists are now able to observe layers of biological and geological activity on the level of the microbial and the subatomic that have never been in the purview of human perception and imagination, opening up new ways of responding to the invisible effects of human-induced violence on a planetary scale. Yet this extended field of sensibilities also raises new ethical challenges. It begs the question: Where do we need to look and how do we need to see to account for deaths and risks of extinction? If the primary basis of our knowledge and relation with the world and with others is no longer confined to the primacy of direct vison and contact, but always already mediated through machine sensors and data proxies, where then do we situate the agency of decision making that is needed to ensure that technical affordances operate in the service of collective betterment and the well-being of all?

At the moment that we become aware of vulnerabilities that transcend prior perceptual registers and scales, the bodies of drowned migrants in the Mediterranean reveal critical fault lines in the governance of risks shared across geological and biological bodies. These fault lines are suggestive of new figural tactics of racialization and abandonment embedded in post-human apparatuses of sense and vision. They encourage us to see how technical innovation extends the calculus of life and death to the full spectrum of energies, frequencies, and signifying practices that the ongoing exchange between social, technical, and biological actors and agencies engenders and that renders individual and collective life chances into a matter of their (im)perceptibility. These differential hierarchies of (im)perceptibility, I suggest, render the new figural tactics of racialization and abandonment visible and apparent. They alert us to the ways technical mediation is implicated in the production and exploitation of "group differentiated vulnerabilities of premature death" which Ruth Wilson-Gilmore famously defined as racism specific to closely interconnected political geographies.[35] This group differentiated exposure to premature death transcends disciplinary boundaries and divisions to conjure a necro-politics that is characteristic for the Anthropocene. It is to the racialized field of (im)perceptibility that I now turn to unpack its necropolitical maneuvers in the historically specific struggle over migration in the Mediterranean Sea.

Sensing Deaths in the Mediterranean

Over the past two years, the European coastguard and border-security agency Frontex has shifted its operational strategy more and more towards intelligence gathering and observational tasks, aimed at identifying and disrupting smuggling networks and other criminal activities at sea. Thus, while search-and-rescue operations are an integral part of Frontex's mandate, it is no longer its first priority, with the result that the vast arsenal of drones, image satellites, vessel-tracking technologies, and sea-, air-, and land-borne radars are primarily used for remote surveillance missions, while the critical task of direct interventions, such as aiding and/or intercepting migrant vessels, is left to humanitarian activists and the national coastguards of South European and North African states.[36] In this scenario Frontex can restrict its monitoring and surveillance activities to the god's-eye view from above, while at the same time strategically preventing migrants from becoming legible to the state by avoiding direct contact, and hence evading potential conditions of accountability.[37]

The skewed system design of Europe's border surveillance did not escape the attention of the European Green Party. On October 10, 2013, the Greens launched a harsh critique in a press release that coincided with the European Parliament's and European Council's endorsement of a legislative agreement on setting up EUROSUR. The press release said:

> The new EUROSUR border surveillance system falls far short of what is needed to save the lives of those who get into difficulty in European waters. EU member states will have to inform Frontex if they are aware of refugees in distress but there is no requirement for them to actually take proactive steps to improve the rescue of shipwrecked refugees by increasing the use of patrol boats in areas that are dangerous for refugees. In addition, they can only request surveillance of the Mediterranean Sea by Frontex for the purpose of preventing "illegal immigration" but not for saving lives.[38]

Hence, according to the Greens, Europe's border-security system misses the main point of protection. Instead of utilizing real-time situational awareness and monitoring capacities for saving lives and supporting migrants in need, surveillance is used to shift responsibility for destitute bodies to countries that lack an asylum system and that may not even be part of the Geneva convention that would at least in principle provide recourse to enforce basic protection and human rights.

One of the biggest challenges in preventing migrant deaths is the fact that irregular border crossings are designed to be untraceable and to escape the radar of border surveillance and the state. Migrants' vessels are not equipped with naval broadcasting systems or transponders, which are required for all passenger and cargo ships, precisely for the purpose of being able to identify their geo-location in case of emergency. This severely complicates securitization as well as the critical tasks of search-and-rescue operations in situations of distress. It leaves migrants by and large dependent on proactive attempts to find them through naval or aerial reconnaissance missions, but also on the commitment of ships roaming across the Mediterranean to adhere to the principles of unconditional assistance, as stipulated in the UN Convention of the Law of the Sea (1982). According to the UN convention, ships have a clear duty to assist those in distress, "regardless of the nationality or status of such persons or the circumstances in which they are found."[39] After the recent upsurge in irregular migration in the Mediterranean, it has become increasingly challenging to enact the law of the sea, not least for commercial shipping companies, who often find themselves inadvertently at the forefront of search-and-rescue operations. As a result, many started to adjust their travel routes to avoid any disruption to global supply chains that depend on the timely delivery of scheduled cargo shipments and reliable travel itineraries. And while states are under a clear obligation to promote the establishment, operation, and maintenance of an adequate and effective search-and-rescue service regarding sea safety, the extent to which these obligations apply to the specific context of irregular migration remains ambivalent, just as the protection afforded to migrants by international human rights law is. As the human rights researcher Stefanie Grant notes: "Although international human rights law protects migrants, it has seldom been applied in situations of border death or loss in the course of migration."[40] As a result, claiming legal responsibility for deaths at sea has proven extremely challenging. To do so, it must first be established that shipmasters, coastguards, or border-security agencies knew or ought to have known of the existence of a real and immediate risk to life, and that they failed to take measures within reasonable limits.[41]

The investigative team Forensic Oceanography has made a rare attempt to prove one such instance of criminal negligence that left sixty-three migrants dead after both military and commercial ships failed to assist them.[42] The case dates back to 2011, when NATO ships regularly patrolled the Libyan coast to enforce an international arms embargo. Military ships, however, do not reveal their geolocation, just like migrant boats, to avoid visibility on open-access vessel tracking systems, such as AIS. AIS stands for Automated Identification System, a real-time monitoring platform that allows the tracking of all registered ships anywhere in the world. Nonetheless, combining AIS data with high-resolution Synthetic Aperture Radar images (SAR), the forensic researchers were able to identify the approximate position of the NATO ships by looking for vessels that were not accounted for by AIS data.[43] One can think of this as a reverse imaging technique that enabled the researchers to subvert the operational logic of tracking technologies. Reading the two data sets against each other in this way provided the evidence needed to show that the NATO ship circled around the migrant vessel several times, but left without assisting them. Such investigative efforts remain the exception, however, because the time and effort required to collect the necessary data are difficult to sustain on a permanent basis, leaving the majority of similar instances unreported or ignored.

Europe's border surveillance system EUROSUR, described earlier, creates similar complications, because access to the information platform is not openly available to investigative reporters or activists. This makes it hard to monitor whether, when, and to what extent Frontex utilizes the vast amount of satellite and ocean monitoring data to rescue or assist migrants in need.

The Politics of Inference; or Seeing through Electromagnetic Waves

The ethical and legal ambivalence surrounding Europe's border surveillance regime in the Mediterranean Sea powerfully underlines how remote-sensing technologies and satellite systems can be used to conceal as much as to reveal what remains otherwise imperceptible or difficult to monitor. What kinds of effects border surveillance systems produce are determined not only by the legal, institutional, or geopolitical contexts in which they are deployed, but also by the performativity of sensory media—their material properties and functionality—in and of itself. As Ballestero notes, sensing technologies never produce a transparent object, readily available for observation, but rather they launch processes of inference, as different wavelengths of the light spectrum touch upon the surface of sensors and inscribe textures on them.[44] Ballestero notes that the "light spectrum is only a tool from which to infer and to gather information about events, that, at the end, are unobservable."[45] Hence, unlike other modes of observation, sensing technologies require multiple acts of interpretation before they yield a recognizable image or representation. In this process, technical instruments and the institutional and social structures behind them converge into "a distributed network of visual-haptic meaning-making" that depends on distinct notions of the marine environment, the tools used to measure and map them, and the concepts people use to make sense of it all.[46] Far from providing a neutral, detached record of the world, free of the human susceptibilities and preoccupations, then, sensory media accomplish a "textural form of knowing"[47] in which the abstract rationalities of scientific templates and models converge with the materiality of touching visions to produce the sea as material witness and archive for evidencing and anticipating risks according to preconceived notions of vulnerability and harm.

Marine biologists have shown great creativity in mobilizing the performativity of Earth-observing media to detect possible threats to the marine ecosystem. A team of researchers at the German Helmholtz Institute just recently developed an algorithm that enables them to identify toxic algal blooms and to assess the effects of global warming on marine plankton, using a distinct spectrum of electromagnetic radiation given off by the sea surface that can be captured by image satellites.[48] Certain groups of phytoplankton can grow to dense masses and produce toxic substances; when there are too many of them in one place, it can be lethal for some marine organisms, especially fish.[49] Using only one aspect of sunlight reflected from the sea to satellite receivers, known as "reflectance,"[50] the scientists were able to identify a unique fingerprint of each of the five known plankton types and to develop an algorithm that can recognize them all. Based on this information they can now produce color-coded maps that show which marine regions are most affected by toxic algae concentrations, and to initiate interventions based on their predictions about the most affected areas.

Similar creativity and inventiveness have been strikingly missing in the governance of irregular migration. This is even more surprising when considering the amount of image satellites and sensor points floating in and above the Mediterranean Sea. The experimental use of these platforms has so far been confined to military research and experiments with oceanographic instruments. This is not only the case in Europe: the US Defense Advanced Research Projects Agency (DARPA), for example, has conducted a series of tests of how aquatic sensors can be mobilized for planetary warfare. The main idea here is to extend the internet of things to poorly connected areas of the high seas by using the sensors to transmit short messages to military stations via satellite.[51]

The fact that the acute need to improve the search-and-rescue capabilities for migrants has so far not been considered in military-scientific experiments points to a constitutive tension at the heart of the applied research and innovation in machine sensing and observation. These technologies, on one hand, opened critical contact zones between humans and nonhumans and their environment as they render what would otherwise be imperceptible visible and apparent. Yet they do not necessarily produce coherent time frames and scales on which this extensive field of sensibilities and awareness is actioned into regimes of protection or care. Rather than nurturing a sense of mutuality and recognition of the interdependencies of risks and vulnerabilities shared across geological and biological bodies, remote-sensing technologies add new layers of complexity to the entangled nature of related being that can be mobilized politically in all sorts of directions and ways.[52] Adrian Lahoud speaks of complexity as a "natural reserve of complication," as an "excess of variables" that open up new possibilities for both oppositional strategies and tactics, and also for new forms of misrecognition, abandonment, and erasure or invisibility. Machine vision, in this sense, incarnates a politics of post-human governance in which accountability, and responsiveness to situations of vulnerability and risk are displaced into the realm of digital textures and data signals that make the eligibility and protection of lives contingent on the ways biophysical properties and behavioral attributes are captured and modeled in data and how they are algorithmically codified.

This displacement of responsibilities is not a systemic glitch or due to institutional failure. It bespeaks a fundamental crisis in the ontologics of post-human governance with regard to the distribution of ethical commitments and solidarities in the face of competing pressures, leaving behind deep inconsistencies in the defense of different kinds of life today and in legitimizations of their protection or abandonment. These inconsistencies are inherently racialized and racializing, creating effects that have yet to be fully acknowledged in the critique of innovation, in particular as it attempts to account for changing constellations of power away from Europe, as the former center of knowledge power, towards a multipolar world.

Concluding Debate

To raise the question of race in relation to technical innovation is not to question the sincerity of techno-scientific practices committed to the critical task of amending the slow violence of colonial-capitalist destruction. Rather it is to stress that this heightened sensibility for the health and survival of oceans and the planet rests on a highly selective recognition of entanglements and interdependencies between humans and nonhumans that may unintentionally redistribute racialized distinctions between natural and premature deaths across species, thereby extending colonial logics deep into the operational scripts of machine sensing and intelligence. Such distinctions, as Singh notes, have historically been used to establish a critical caesura between populations that depreciate one form of humanity for the purpose of another's health, development, and safety, predisposing them to "group-differentiated vulnerabilities of pre-mature death."[53] Now that the risk of premature death reveals itself as ecological crises on a planetary scale, it is imperative to understand how such racialized determinations articulate the extended field of sensibility afforded by machine sensing and vision. Not least because remote viewing technologies have become the primary interface for rendering the oceans, environments, and the Earth readable, knowable, and addressable, without them it will be impossible to fathom, much less to contain, threats of extinction and death.

What is more, as biophysical and geological processes are becoming ever more central to the question of life and the conditions of its ending, the vulnerabilities and exposure to premature deaths are no longer reducible to an effect of extralegal or state power, much less to the power over human life as such. Rather they need to be considered through the fundamental challenge of how to reconcile the singularity of individual death in the face of total death as it announces itself in the figure of the Anthropocene. As Mbembe notes, "For a long time we have been concerned with how life emerges and the conditions of its evolution. The key question today is . . . under what conditions it ends . . . or how it can be repaired, reproduced, sustained and cared for, made durable, preserved and universally shared."[54] By evoking this question of how life ends and the concepts and tools through which these endings are made legible and recognizable, the bodies of dead migrants present us with a critical limit for thinking innovation in a multipolar world. These deaths remind us of the racialized predicates on which the idea of universal betterment has built and that continue to provide one of the central tenets of technological advances in the name of sustainable futures and development. They encourage us to remain cautious about the ways racial fictions project their normative horizon onto the extended field of sensibilities for knowing, seeing, and anticipating life and deaths today.

The ways in which migrants in distress have remained by and large invisible in the partial renderings of the Mediterranean Sea calls for a rethinking of race, to make room for acknowledging race as technology rather than as skin color or biophysical essence. As Singh suggests, we need to recognize the technology of race as "precisely those historic repertoires and cultural, spatial, and signifying systems that stigmatize and depreciate one form of humanity" for the sake of another's health, development, safety, profit, and pleasure.[55] Environmental sensing and Earth observation technologies, I suggest, need to be considered as a critical component of these signifying practices and systems. They facilitate a racialized politics of (im)perceptibility, whose exclusionary tactics, to follow Yusoff, have "a critical bearing on the co-habitation of worlds."[56] Recognizing the potential violence of (im)perceptibility is of particular importance at this historical juncture, when the distinctions between military and environmental infrastructures are becoming ever more difficult to ascertain, and their historical entanglements are ever more difficult to ignore. These entanglements unfold against a cultural backdrop that is deeply grounded in Cartesian divisions between science and politics, the laws of nature and the laws of the state. Such divisions allow for a selective recognition of risks through technical mediation, whose particular mode of dehumanization manifests itself not only in the structural invisibility and misrecognition of certain bodies, but in the ways their deaths and disappearance are left unspecified and nondescript.

Allen Feldman speaks of such targeted elisions as "containerization of war" to describe the perceptual and juridical blurring of military and scientific intelligence in digital platforms and warfare.[57] Building on the work of Allan Sekula, containerization here denotes the offshoring of violence in which war becomes "the mode by which executive power is implanted and expands through the perceptual scattering" of situational awareness.[58] The increasing convergence of military and scientific platforms for maritime surveillance are key drivers of this diffraction and outsourcing of violence into "rarefied channels of computation."[59] They facilitate "select scenic affirmations" of (in)visible threats on the basis of objectifying measures of biophysical and biochemical properties and relations, such as rising sea temperatures, salinity levels, or toxic algae concentrations, that diffuse and diffract racialized state power into the performativity of data textures and, ultimately, into the materiality of communication itself. This is a form of infrastructural war that does not work on objects or bodies directly; rather it operates through techno-scientific renderings and tactical elisions that "declassify certain bodies and populations as proper to the terrestrial surface," thereby denying them "political exemplification" and intelligibility.[60] Relayed back to the question of historical continuities in a more-than-human multipolar world, such "acts of vanishment"[61] bring into relief how the planetary infrastructures for governing life and death implicate bio/geological processes and relations in colonial logics that extend the idea of the border deep into the new contact zones of nature-cultures Earth observation technologies and sensing devices created. They conjure an expansive field of sensibilities that allows for some deaths to be naturalized, while nature is historicized into a material witness and archive for determining what counts as risk, and in the name of who or what it is confronted within regimes of protection or care.

NOTES

1. World Wildlife Fund, "Exploring How Climate Change Relates to Oceans," 2021, https://www.worldwildlife.org/stories/how-climate-change-relates-to-oceans.
2. Sandra Sendra et al., "Oceanographic Multisensor Buoy Based on Low Cost Sensors for Posidonia Meadows Monitoring in Mediterranean Sea," *Journal of Sensors* 2015 (2015), doi: 10.1155/2015/920168.
3. Jennifer Gabrys, "Ocean Sensing and Navigating the End of This World," 4, https://www.e-flux.com/journal/101/272633/ocean-sensing-and-navigating-the-end-of-this-world/.
4. The European Parliament and the Council of the European Union, *Section 3, Article 18 Eurosur: REGULATION (EU) 2019/1896 the EUROPEAN PARLIAMENT and the COUNCIL on the European Border and Coast Guard and Repealing Regulations (EU) No 1052/2013 and (EU) 2016/1624* (Brussels: The European Parliament and the European Union, 14 November 2019), 19–23; Chris Jones, "Monitoring 'Secondary Movements' and 'Hotspots': Frontex Is Now an Internal Surveillance Agency," https://www.statewatch.org/media/documents/analyses/no-348-frontex-internal-surveillance.pdf.
5. The European Parliament and the Council of the European Union, *Section 3, Article 18 Eurosur: REGULATION (EU) 2019/1896 the EUROPEAN PARLIAMENT and the COUNCIL on the European Border and Coast Guard and Repealing Regulations (EU) No 1052/2013 and (EU) 2016/1624*.
6. Asia Maritime Transparency Initiative, "A Survey of Marine Research Vessels in the Indo-Pacific | Asia Maritime Transparency Initiative," https://amti.csis.org/a-survey-of-marine-research-vessels-in-the-indo-pacific/; Asia Maritime Transparency Initiative, "Exploring China's Unmanned Ocean Network | Asia Maritime Transparency Initiative," https://amti.csis.org/exploring-chinas-unmanned-ocean-network/.
7. Jennifer Gabrys, *Program Earth: Environmental Sensing Technology and the Making of a Computational Planet* (Minneapolis: University of Minnesota Press, 2016), 12.

8. Paolo G. Albano et al., "Historical Ecology of a Biological Invasion: The Interplay of Eutrophication and Pollution Determines Time Lags in Establishment and Detection," *Biological Invasions* 20, no. 6 (2018), doi: 10.1007/s10530-017-1634-7; Bella S. Galil et al., "'Double Trouble': The Expansion of the Suez Canal and Marine Bioinvasions in the Mediterranean Sea," *Biological Invasions* 17, no. 4 (2015), doi: 10.1007/s10530-014-0778-y.
9. IOM, "Missing Migrants: Tracking Deaths along Migratory Routes," IOM, https://missingmigrants.iom.int/.
10. Kathryn Yusoff, *A Billion Black Anthropocenes or None*, Kindle, Forerunners: Ideas First from the University of Minnesota Press [53] (Minneapolis: University of Minnesota Press, 2018), https://manifold.umn.edu; Nigel Clark, "Politics of Strata," *Theory, Culture & Society* 34, no. 2–3 (2017); Richard A. Grusin, ed., *Anthropocene Feminism*, 21st Century Studies (Minneapolis: University of Minnesota Press, 2017).
11. Monika Halkort, "Dying in the Technosphere: An Intersectional Analysis of European Migration Maps," in *Mapping Crisis: Participation, Datafication and Humanitarianism in the Age of Digital Mapping*, ed. Doug Specht (London: Institute of Commonwealth Studies, 2020).
12. See Hoyng and Chong in the introduction of this volume.
13. Yusoff, *A Billion Black Anthropocenes or None*; Françoise Verges, "Racial Capitalocene: Is the Anthropocene Racial?," https://www.are.na/block/3428729; Roberto Mantovani, "The Otranto-Valona Cable and the Origins of Submarine Telegraphy in Italy," *Advances in Historical Studies* 6, no. 1 (2017), doi: 10.4236/ahs.2017.61002.
14. Helen Tilley, *Africa as Living Laboratory: Empire, Development, and the Problem of Scientific Knowledge, 1870–1950* (Chicago: University of Chicago Press, 2011).
15. Kathryn Yusoff, "White Utopia/Black Inferno: Life on a Geologic Spike," *e-flux*, 2019, https://www.e-flux.com/journal/97/252226/white-utopia-black-inferno-life-on-a-geologic-spike/; Angelo Matteo Caglioti, *The Climate of Fascism: Science, Environment, and Empire in Liberal and Fascist Italy (1860–1960): Ongoing Project* (Rome: American Academy in Rome, 2019); Martin Mahony and Georgina Endfield, "Climate and Colonialism," *Wiley Interdisciplinary Reviews: Climate Change* 9, no. 2 (2018); David N. Livingstone, "Race, Space and Moral Climatology: Notes toward a Genealogy,'" *Journal of Historical Geography* 28, no. 2 (2002): 2.
16. Achille Mbembe, interview by Torbjørn Tumyr Nilsen, November 30, 2018, Bergen, Norway; Yusoff, *A Billion Black Anthropocenes or None*.
17. Yusoff, *A Billion Black Anthropocenes or None*, loc 217;225.
18. Valeska Huber, "Connecting Colonial Seas: The 'International Colonisation' of Port Said

and the Suez Canal during and after the First World War," *European Review of History: Revue européenne d'histoire* 19, no. 1 (2012): 144.

19. Paolo G. Albano, "Historical Ecology of Lessepsian Migration," https://www.univie.ac.at/lessepsian/.
20. Huber, "Connecting Colonial Seas," 143.
21. Huber, "Connecting Colonial Seas," 143.
22. Galil et al., "'Double Trouble': The Expansion of the Suez Canal"; Albano, "Historical Ecology of Lessepsian Migration."
23. Callum Roberts, *The Ocean of Life: The Fate of Men and the Sea* (New York: Viking Penguin, 2012), 179.
24. Roberts, *The Ocean of Life*, 181.
25. Rafting debris, such as ocean buoys and marine litter, are other convenient ways for nonindigenous species to travel long distances and to exploit ecological niches; Angelina Ivkić, Jan Steger, Bella Galil, and Paolo Albano, "The Potential of Large Rafting Objects to Spread," *Biological Invasions* 21, no. 6 (*2019*).
26. Roberts, *The Ocean of Life*, 183.
27. Kathryn Yusoff, "Geologic Life: Prehistory, Climate, Futures in the Anthropocene," *Environment and Planning D: Society and Space* (2013): 5.
28. Yusoff, "Geologic Life."
29. Yusoff, "Geologic Life," 781.
30. Jessica Lehman, "A Sea of Potential: The Politics of Global Ocean Observations," *Political Geology* 55 (2016): 121.
31. Lehman, "A Sea of Potential," 120.
32. Stefan Helmreich, *Alien Ocean: Anthropological Voyages in Microbial Seas* (Berkeley: University of California Press, 2009).
33. Helmreich, *Alien Ocean*; David Chandler, "Mapping beyond the Human: Correlation and the Governance of Effects," in *Mapping and Politics in the Digital Age*, ed. Pol Bargués-Pedreny, David Chandler, and Elena Simon (London: Routledge, 2018); Lehman, "A Sea of Potential."
34. Ben Anderson, "Preemption, Precaution, Preparedness: Anticipatory Action and Future Geographies," *Progress in Human Geography* 34, no. 6 (2010): 779; doi: 10.1177/0309132510362600.
35. Ruth Wilson Gilmore, *Golden Gulag: Prisons, Surplus and Opposition in Globalizing California* (Berkeley: University of California Press, 2007), 28.
36. Maurice Stierl, interview by Monika Halkort, Rome, 2019.
37. Mark Latonero and Paula Kift, "On Digital Passages and Borders: Refugees and the New

Infrastructure for Movement and Control," *Social Media + Society* (2018); Statewatch, "EU: Frontex Splashes Out: Millions of Euros for New Technology and Equipment," 2018, https://www.statewatch.org; Alarm Phone et al., "Remote Control: The EU-Libya Collaboration in Mass Interceptions of Migrants in the Central Mediterranean," June 17, 2020, https://eu-libya.info/img/RemoteControl_Report_0620.pdf.
38. Ska Keller, "European Border Surveillance (EUROSUR)," https://www.greens-efa.eu/en/article/press/european-border-surveillance-eurosur-4777/.
39. Norton Rose Fulbright, "The Rescue of Migrants at Sea—Obligations of the Shipping Industry," https://www.nortonrosefulbright.com/en/knowledge/publications/09f857fc/the-rescue-of-migrants-at-sea---obligations-of-the-shipping-industry.
40. Stefanie Grant, "Dead and Missing Migrants: The Obligations of European States under International Human Rights Law: IHRL Briefing" (2016), 8–10, http://iosifkovras.com/wp-content/uploads/2015/08/Legal-Memo-on-Missing-Migrants.pdf. According to Grant, the state is under a positive—substantive—duty to take preventive action where there are foreseeable threats to life originating not only from state authorities or private persons but also from environmental hazards or self-induced risks that result from an individual's own actions. And while this would imply that states should not exclude migrants who pay smugglers to take them on dangerous journeys from these duties, the specific context of irregular migrants has not yet been considered by the European Court of Human rights (ECHR).
41. Grant, "Dead and Missing Migrants," 11.
42. Charles Heller and Lorenzo Pezzani, "The Left-to-Die-Boat," https://forensic-architecture.org/investigation/the-left-to-die-boat.
43. Heller and Pezzani, "The Left-to-Die-Boat."
44. Andrea Ballestero, "Touching with Light, or, How Texture Recasts the Sensing of Underground Water," *Science, Technology & Human Values* 44, no. 5 (2019): 14.
45. Ballestero, "Touching with Light," 16.
46. Ballestero, "Touching with Light," 15, 17.
47. Ballestero, "Touching with Light," 17.
48. Science Daily, "Observing Phytoplankton via Satellite," *Science Daily*, https://www.sciencedaily.com/releases/2020/03/200319125151.htm.
49. Science Daily, "Observing Phytoplankton via Satellite."
50. Reflectance (or coefficient of reflection) represents the amount of sunlight striking the Earth that is reflected back into space and that can be measured by satellite sensors.
51. George Seffers, "Afcea.Org: DARPA's Ocean of Things Ripples across Research Areas," https://www.afcea.org/content/darpas-ocean-of-things-ripples-across-research-areas.

52. Adrian Lahoud, "Geo-Politics: Conflict and Resistance in the Anthropocene," Anthropocene Campus, https://www.youtube.com/watch?v=f7UquyLLTYY, min 03.44.
53. Cited in Françoise Verges, "Racial Capitalocene: Is the Anthropocene Racial?" In *Futures of Black Radicalism*, ed. Gaye Theresa Johnson and Alex Lubin (New York: Verso, 2017), 10.
54. Achille Mbembe, interview by Torbjørn Tumyr Nilsen.
55. Cited in Verges, "Racial Capitalocene," 10.
56. Kathryn Yusoff, "Insensible Worlds: Postrelational Ethics, Indeterminacy and the (K) Nots of Relating," *Environment and Planning D: Society and Space* 31, no. 2 (2013).
57. Asif Akthar and Selim Karlitekin, "On War, Photopolitics, White Public Space and the Body: A Conversation with Allen Feldman," *Nakedpunch*, http://www.nakedpunch.com/articles/273; Allan Sekula, "Freeway to China," *Public Culture* 12, no. 2 (2000): 2.
58. The shipping container, in Sekula, "Freeway to China," is the very coffin of remote labor power, bearing the hidden evidence of exploitation in the far reaches of the world. It enables corporations to offshore the violence of extractive labor relations and to escape legal accountability while intensifying their predatory practices.
59. Akthar and Karlitekin, "On War, Photopolitics, White Public Space and the Body."
60. Allen Feldman, "War under Erasure: Contretemps, Disappearance, Anthropophagy, Survivance," *Theory & Event* 22, no. 1 (2019): 186.
61. Feldman, "War under Erasure."

Conclusion

Futures in the Plural

Jack Linchuan Qiu

The project behind this book started at the National Communication Association's Hong Kong Communication Workshop (https://bit.ly/3hl1EcO), held on June 28–29, 2019, at the Centre for Chinese Media and Comparative Communication Research (C-Centre), on the lovely campus of the Chinese University of Hong Kong (CUHK). The workshop was held in Hong Kong because some of the participants in the larger gathering known as the Shenzhen Forum were disinvited by the authorities in Mainland China. Participants therefore referred in a ridiculing manner to this workshop as a "salon des refusés." It was a special honor for C-Centre to host these self-acclaimed "rejects." The workshop was a great success thanks to the masterful organizing by Professor Rolien Hoyng, who edits this volume along with Professor Gladys Chong based on papers from the Hong Kong Workshop. The result is a systemic and seminal treatment of some of the most pressing issues facing media and communication research: innovations, platforms, infrastructures that are technological as well as social and environmental, and comparative analysis in the contexts of rapidly changing global and regional geopolitics.

The world is in deep disarray at the beginning of the new decade of the 2020s. It's precisely in such moments when we need to rethink our fundamental assumptions, reexamine our methods and approaches, reorganize our materials and analyses, and reimagine the futures of communication studies. How to make sense of this world, now characterized by Big Tech hegemony, infrastructure breakdowns, surging xenophobia, and myriad forms of resistance and creativity from the bottom up? How to compare the various platforms, digital cultures and subcultures, and models of internet policy, beyond the supposedly default centers of high-tech geography, be they Californian or Chinese?

Critical and interdisciplinary, chapters in this book draw from a wide spectrum of scholarly traditions: media and communication research, critical political economy, cultural studies, legal studies, geography, international development, and more. Each within its own contexts, asking different research questions and making unique contributions, all chapters converge in their analytical focus on new information and communication technologies (ICTs); their inner operational logics, especially surrounding the question of innovation; and their broader implications for the media industries, global economy, society at large, and Planet Earth as a whole. The topics covered include classic issues such as capital and labor, media globalization, and the de-westernization of communication research, as well as cutting-edge work on digital platforms, internet governance, technology design, making/makers, and the posthuman consequences of the Anthropocene.

Hoyng and Chong structure the chapters into three parts: (a) *formal innovations* through datafication, financialization, and reterritorialized high-tech development in China, Europe, and Turkey—chapters 1–3; (b) *everyday inventiveness* through the practices of makers, smartphone designers, and short video platforms—chapters 4–5; and (c) *novelty as technodiversity* captured through design processes seeking to emulate "indigenous innovation" and the intended and unintended consequences of ocean-seeing technologies—chapters 6–7. This potent framework is not about three discrete categories of things that stand parallel to each other. Rather, the formal often interact with the informal, everyday practices, while both the formal and informal produce, and are shaped by, bio-social-technical conditions that are not only cultural and discursive but also natural and environmental.

As such, Hoyng and Chong's tripartite conceptualization offers a dynamic opportunity for synthesis, on the basis of which we can posit about the futures—in the plural—of media and communication research in the 2020s. More specifically, we can—and must—now reconsider at least three fundamental questions about (1) the dialectics of the new, (2) comparative communicative research, and (3) the meanings of a multipolar world.

Dialectics of the New

Researchers of ICTs and media phenomena have long chased cutting-edge innovations, the latest popular brands, trending concepts, methods, memes, and hashtags. But what actually constitutes novelty and creativity? Why do new things deserve so much scholarly scrutiny—at the expense of the past? Are we fetishizing the new, not because they have any intrinsic value, but because our academic agenda is set by the tech and media corporations that are under the pressure of Wall Street quarterly reports and the capitalist logic of expansion and "planned obsolescence"?[1] Or, is it because we simply want to run away from old, unresolved problems, for which chasing running targets would serve as an excuse for incompetence? Chapters in this volume invite us to ponder these hard questions.

To define what's new, we must first establish what's old. Novelty and normalcy are relative. Historicizing the subject matter is indispensable for effective theory-building. Coincidentally, when the NCA Hong Kong Communication Workshop took place, the event venue had the motto of the CUHK School of Journalism and Communication displayed right outside C-Centre, which read: "Inherit, Innovate, Inspire." *Inherit* means that, before engaging the new, we must first take stock of the old while curating existing traditions. Otherwise, we cannot really innovate. Without heritage and pedigree, the innovations, even if materialized, will probably be superficial, myopic, ephemeral. *Innovate* is to break new paths that simultaneously address classic questions while leading to durable advancements in knowledge production, so that the new paths broken will be walked upon, maintained, and extended by generations of young scholars. As such, real innovations are way more than cool gadgets for display, to cater to popular tastes or appeal to the latest vogue. *Inspire* signifies the extraordinary capacity of true innovations, building on inheritances but not constrained by them, to spur knowledge-making dynamism out of the box, defying gravity and conformity.

The studies in this book, each in its own way, inherit, innovate, and inspire us to transcend the received wisdom—and exclusive fixation—on novelties, often defined conveniently and superficially. One such fetish is Silicon Valley, complete with its "fever,"[2] rapidly expanding corporations, neoliberal entrepreneurialism, and alpha-male "new rich," now expanding to the other side of the Pacific such as Beijing's Zhongguancun District, aka "the Silicon Valley of China." How many more "Silicon Valleys" do we need to endlessly reinforce the "Californian ideology" and "old colonial philosophy" of Eurocentrism, including in the high-tech showrooms of Shenzhen?[3]

Another indicator of prevailing intellectual laziness is for scholars to pick up popular buzzwords, legitimize and capitalize on them, while blindly following the agenda of the powers that be. The jargon can be BRICS or emerging markets, blockchain or quantum computing. One of the most popular terms is "platform," which originated from the Google–YouTube merger in 2008.[4] Yet, is platform really the right concept?[5] Or, after unpacking the political economy and discursive power behind it, would it be better to use the old term of infrastructure? It's oftentimes more productive, and more fundamentally innovative, to deconstruct such new terminologies from the corporate world through long-standing traditions of critique—for example, by following the political economy tradition of analyzing (new) media systems in connection with financial capital (Jia and Nieborg, chapter 1).

Similarly, in addition to the obsession with Silicon Valley and American tech giants, a renewed version of techno-orientalism is becoming prevalent regarding Chinese unicorn companies such as Alibaba and Tencent. It's renewed because the genre was well-established since the 1980s when Japanese corporations and tech culture were worshiped and fetishized in the West. But does this "rise of China" since the turn of the century mean something genuinely new, above and beyond "Japan as No. 1" decades ago? Is it merely another instance of techno-orientalism? Or, as Fujiwara and Nagano suggest, is Japan becoming a part of "America's informal empire,"[6] along with the Philippines? While Japan and the Philippines represent two pieces of Americana in the informal juggernaut of the United States in the Asia-Pacific, China can be seen as comprising both high- and low-end extensions at the same time—for example, as seen in chapters 4 to 6 of this volume.

Techno-orientalism, be it Chinese or Japanese, may seem like a new force that challenges the old US-centric order, at least in geographical space. But more importantly, it reinforces the stereotype of the East as the Other, the miraculous, the exotic. It conceals the undercurrents of the informal empire such as financialization that may not follow the fault lines of geopolitics (Jia and Nieborg, chapter 1). Such a view also marginalizes other players, especially the EU, whose GDPR deserves in-depth analysis, especially if the futures at stake are not just neoliberal but can also be "postcapitalist" (Daly, chapter 2).

The new shape the old, and are shaped by the old in turn, whether it's new technologies, new players, new questions, or new ways of thinking. This edited volume reminds us about the dialectics between tradition and innovation, the need to critique the new as the latest trending phenomenon being wrapped within capitalism, Eurocentrism, and techno-orientalism. Some of these fetishized novelties can be extremely seductive, not because they break any new ground, but because they are phantoms of the past.

Comparative Perspectives

How can we study the infrastructures of communication innovation critically? Answering this question requires comparative analysis. The field of media and communication studies, like many other fields, has always been implicitly comparative, although explicit engagement with issues of comparative methods was a more recent development after post–World War II decolonialization and efforts of nation-building and industrialization that relied on the establishment of national print and broadcast media systems in the newly independent nations. This was a politicized period for the design and buildup of communication infrastructures, which reached its peak at the turning point of the 1980s marked by UNESCO's *Many Voices, One World* as well as the "open skies policy" of the United States.[7] Meanwhile, comparative communication research entered the limelight with the end of the Cold War, borrowing from more established fields such as comparative politics.[8] Exemplary works from this period include trans-Atlantic election coverage studies that led to such key texts as Hallin and Mancini's *Comparing Media Systems*.[9]

The world is now very different as we enter the 2020s. Digital transformations of our time depend much less on sovereign states and supranational organizations such as the UN, much more on private companies, transnational capital, and networks of nonstate actors. Google and Al-Qaeda are, in this sense, both "communication innovators" that operate across national borders. So are makers in Shenzhen (Mutibwa and Xia, chapter 4), game companies in Turkey (Sezgin and Binark, chapter 3), and smartphone brands in Ghana (Lu, chapter 6), not to mention major social media platforms: Facebook, TikTok, YouTube, etc. These privately owned, transnational entities have become the backbone of digital communication, globally and regionally. They are the contemporary equivalents of national print and broadcast systems of the previous era, yet most of them are not national anymore.

However, it is erroneous to see corporate platforms as complete substitutes for national media systems. Still, there are Wikipedia, the Free and Open Source Software (FOSS) movement, and what Mutibwa and Xia terms "countercultural values" (chapter 4) that may harbor anti- and postcapitalism. While the balance of power has tilted away from nation-states, some of them—the more resourceful ones—have stepped up control over tech giants through new measures such as the GDPR (Daly, chapter 2) and anti-trust investigations against the likes of Google and Alibaba (https://bit.ly/2XtrRg2). This happens while the unicorn companies start to clash with and weaken each other in an unprecedented manner, no matter if it's Facebook against Apple or Huawei against Tencent. Meanwhile, new surveillance and data processing technologies open a new chapter of "posthuman governance," for instance in the Mediterranean Sea (Halkort, chapter 7), where the socio-technical has come to hinge ever more increasingly on the geobiological as the world's environmental crisis and the predicaments of migrant refugees deteriorate.

What on earth should we compare, using what units of analysis? Nations are still relevant, but no longer exclusively so. Chapters in this volume have examined digital platforms (e.g., chapters 1, 5), communities of makers and designers (chapters 4 and 6), and global and regional communication infrastructures in the forms of the ocean and biomass (chapter 7). At this time of existential crises, a sustainable Planet Earth with its climate stability and biodiversity is the ultimate infrastructure that needs care and renewal. The global unit of analysis should in this sense encompass other life forms and the physical environment beyond the Anthropocene.[10]

In comparing these various media and communication systems—be they local, national, regional, global, and/or planetary—chapters in this book avoid the tendency to see everything through the angle of Silicon Valley, as if innovations can only be appreciated in the shadow of the United States. Chinese smartphone companies operating in Africa are ahead of their competitors from the Global North in achieving more "design justice" beyond (neo)colonialism precisely because they do not use Apple as a benchmark (Lu, chapter 6).[11] To make sense of game companies operating in the oppressive environment of Turkey, Sezgin and Binark (chapter 3) refer to China and its dissident artist Ai Weiwei. Such inter-Asian and South-South comparison can often be more productive than conventional US-centered frameworks.

Beyond a Uni- or Bipolar World

Decentering the old, unhinging it from intellectual pedestals, is but a first step toward fully appreciating the new. What happens if Silicon Valley ceases to be the only center for the world of digital communication innovations? Too often the narrative would turn to China, for good or for bad, as the competing force against the United States in not only geopolitics but also technoculture and R&D (Jia and Nieborg, chapter 1). While the US model of communication infrastructure is neoliberal and global, the Chinese model presumes the territorial basis of a sovereign nation-state controlling media and tech companies in pursuit of economic profits while maintaining the domestic political-economy status quo. Monroe Price contrasts the two as "vampires and ghosts,"[12] suggesting that while they have different propensities and appetites, they are also fundamentally similar creatures whose scope of activities are limited. This suggests that the conception of a bipolar world does not do justice to the full spectrum of possibilities for institutional formation. It is rather insufficient and may distract us from understanding future possibilities.

The poverty of dualistic conceptions extends from the persistence of Cold War mentalities. The first mistake is to see everything through the lens of China vis-à-vis the United States, either as China-embracing or China-bashing. What about the continued trans-Pacific ties between American and Chinese companies in R&D, finance, and tech movements, as addressed in chapters 1 and 4? Even if politicians' efforts in cutting such ties are successful, it is likely that Chinese high-tech unicorns will continue operating under the pressure of Hong Kong, Shanghai, and Shenzhen Stock Exchanges in such typical "American" ways without any US company or individual partaking directly in the game. Ultimately, the supposedly different paths may lead to the same destination of a digital future under the auspices of financial capital, be it American-style or with Chinese characteristics. To get there, a fateful clash of empires would be inevitable, and the only difference we can make is to take a side, minimize the chance of mutual annihilation, and ensure "our side" will win.

There is nothing new in such a seemingly bipolar but ultimately converging mindset, which is imperialist in nature. Chapters in this volume have demonstrated ways to transcend and/or disavow it by focusing on other parts of the world, be they Europe, Turkey, or the Mediterranean Sea (chapters 2, 3, and 7). Although it's impossible to be all-inclusive in one volume, it's important to stress that significant clusters of communication infrastructure innovations have also been emerging in Korea, Japan, Taiwan, India, Israel, Kenya, Russia, South Africa, Brazil, and more. Players in these societies include statist players in government-owned entities such as the military, as well as civic tech and activist groups struggling for social and environmental justice, including data justice, through the redesign and reimagination of digital infrastructures.

Appreciating a multipolar world thus requires a much broader view to examine more regions of the world and to embrace more diversified types of players operating beyond and beneath the Westphalian framework centered on sovereign nation-states. This shall allow us to see the failures of American expansionism in other parts of the world as well as domestically, through such events as Edward Snowden. The same can be seen much more frequently in Chinese companies failing to achieve dominance or even survive overseas, not to mention the downfall of "Made in China 2025."[3] Indeed, as critical scholars we must see through the hubris of discursive oversimplification and base our argument on solid empirical evidence. We must deconstruct lingering US-centrism, rising Chinese exceptionalism, and techno-orientalism in an effort to make sense of the multipolar world, hence envisioning our collective digital futures as they grow not only from corporate boardrooms and elite university labs but also from the unlikely places of the Global South. These innovative futures can be market- or nonmarket-based, involving human or other actors, responding to existential needs of civic organizations, the disenfranchised, and the voiceless, including the bio-socio-technical infrastructures of Planet Earth. This is how we approach genuine multipolarism and our digital futures—in the plural, and working from the premise of solidarity.

NOTES

1. Dallas Walker Smythe, *Counterclockwise* (Boulder, CO: Westview Press: 1994).
2. Everett M. Rogers and Judith K. Larsen, *Silicon Valley Fever: Growth of High-Technology Culture* (New York: Basic Books, 1984).
3. Richard Barbrook and Andy Cameron, "The Californian Ideology," *Science as Culture* 6, no. 1 (1996): 44–72.
4. Tarleton Gillespie, "The Politics of 'Platforms,'" *New Media & Society* 12, no. 3 (2010): 347–64.
5. Guobin Yang and Wang Wei, eds., *Social Media in China: Platforms, Publics, and Production* (East Lansing: Michigan State University Press, 2021).
6. Kiichi Fujiwara and Yoshiko Nagano, *America's Informal Empires: Philippines and Japan* (Manila: Anvil Publishing, 2012).
7. Sean MacBride and Elie Abel, *Many Voices, One World: Communication and Society, Today and Tomorrow* (Paris: UNESCO, 1984). Christopher H. Sterling, Phyllis Bernt, and Martin B. H. Weiss, *Shaping American Telecommunications: A History of Technology, Policy, and Economics* (New York: Routledge, 2006).

8. Jay G. Blumler, Jack M. Mcleod, and Karl Erik Rosengren, eds., *Comparatively Speaking: Communication and Culture across Space and Time* (Newbury Park, CA: Sage, 1992).
9. Daniel C. Hallin and Paolo Mancini, *Comparing Media Systems: Three Models of Media and Politics* (Cambridge: Cambridge University Press, 2004).
10. Jingfang Liu and Phaedra C. Pezzullo, eds., *Green Communication in China: On Crisis, Care, and Global Futures* (East Lansing: Michigan State University Press, 2021).
11. Sasha Costanza-Chock, *Design Justice: Community-Led Practices to Build the Worlds We Need* (Cambridge, MA: MIT Press, 2020).
12. Monroe E. Price, "Ghosts, Vampires, and the Global Shaping of Internet Policy," in *Law and Disciplinarity*, ed. Robert J. Beck (New York: Palgrave Macmillan, 2013), 229–44.
13. Issaku Harada, "Beijing Drops 'Made in China 2025' from Government Report," *Nikkei Asia Review*, March 6, 2019, https://asia.nikkei.com/Politics/China-People-s-Congress/Beijing-drops-Made-in-China-2025-from-government-report.

Contributors

Mutlu Binark is Professor at the Department of Radio-Television and Cinema, and chair of the Division of Informatics and Information Technologies at the Faculty of Communication, Hacettepe University. She has been the editor of *Moment Journal* since July 2017. She is also a founding member of Alternative Informatics Association. She teaches media theories, media sociology, media literacy, and new media culture. She is currently working on aging, data ethics, creative content industries, and cultural policies both in Turkey and Asia. Recent publications include a chapter in *Authoritarian and Populist Influences in the New Media* (with G. Bayraktutan, 2018), a book, *Cultural Diplomacy and Korean Wave "Hallyu": K-movies, K-dramas and K-pop* (2019), and an edited book and chapters in *Popular Culture and Media in Asia* (2020). Her blog is www.yenimedya.wordpress.com.

Gladys Pak Lei Chong is Associate Professor of the Department of Humanities and Creative Writing at Hong Kong Baptist University. Her research interests include Chinese governmentality, cultural governance, power relations, process of subjectification, discourse analysis, gender, place making, social credit, security, surveillance, and technology. She is the author of *Chinese Subjectivities and the Beijing Olympics* (2017) and coeditor of *Trans-Asia as Method: Theory and Practices* (2020). Some of her recently published journal articles have appeared in *The Information Society*,

Chinese Journal of Communication, *Journal of Current Chinese Affairs*, and *China Information*. Her research on youth aspirations, and technology, security, and risk is funded by the Hong Kong Research Grant Council.

Angela Daly is Professor of Law & Technology at the University of Dundee (Scotland). She is a socio-legal scholar of the regulation and governance of new (digital) technologies. She is the author of *Socio-Legal Aspects of the 3D Printing Revolution* (2016), and *Private Power, Online Information Flows and EU Law: Mind the Gap* (2016) based on her doctoral research at the European University Institute. She is also coeditor of the open-access collection *Good Data* (2019). She previously worked at the University of Strathclyde Law School (Scotland), the Chinese University of Hong Kong, Queensland University of Technology, and Swinburne University of Technology (Australia).

Jeroen de Kloet is Professor of Globalisation Studies and Head of the Department of Media Studies at the University of Amsterdam. He is also a professor at the State Key Lab of Media Convergence and Communication, Communication University of China in Beijing. Publications include a book with Anthony Fung, *Youth Cultures in China* (2017), and the edited volumes *Boredom, Shanzhai, and Digitization in the Time of Creative China* (with Yiu Fai Chow and Lena Scheen, 2019), and *Trans-Asia as Method: Theory and Practices* (with Yiu Fai Chow and Gladys Pak Lei Chong, 2019). See also http://jeroendekloet.nl.

Monika Halkort currently lectures at the University of Applied Arts in Vienna, Austria. From 2013 to 2020 she was Assistant Professor of Digital Media and Social Communication at the Lebanese American University (LAU) in Beirut, Lebanon. Monika researches and writes about the political and moral ecology of digital infrastructures, data colonialism, and the material entanglement of techno-scientific instrument platforms with racialized knowledge regimes. Her work has appeared in peer-reviewed academic journals such as the *International Journal for Communication*, the *Canadian Journal of Communication*, and *Tecnoscienza* as well as edited books, including *Mapping Crisis: Participation, Datafication and Humanitarianism in the Age of Digital Mapping*, edited by Doug Specht (2020) and *Oceans Rising*, edited by Daniela Zyman and Markus Reyman (2021). The main geographic focus of her work is the Arab world and the Mediterranean South.

Rolien Hoyng is Assistant Professor at the School of Journalism and Communication of The Chinese University of Hong Kong and director of the Master in Global Communication. Her research addresses digital infrastructures and data-centric technologies in particular contexts of practice, including urban politics and dissent as well as ecological politics and electronic waste. Moreover, her multi-sited research encompasses Turkey, Hong Kong, and Europe, which prompts her interest in comparative possibilities. Recently published work has appeared in *Antipode, Television & New Media*, and *International Journal of Cultural Studies*. She is the editor of a special issue entitled "Digital Infrastructure, Liminality and World-Making via Asia," published with the *International Journal of Communication*.

Lianrui Jia is a Lecturer in Digital Media and Society in the Department of Sociological Studies at the University of Sheffield. She holds a PhD in Communication Studies from York University. Her postdoctoral research, funded by Canada's Social Sciences and Humanities Research Council, examined the rise of Chinese digital platforms, the app economy, and their global impacts. Her research on Chinese internet governance, political economy, and globalization has been published in *International Communication Gazette, Internet Policy Review, European Journal for Cultural Studies, Internet Histories*, and *Westminster Papers in Communication and Culture*.

Jian Lin is an Assistant Professor in the Department of Media and Journalism Studies at the University of Groningen. He has coauthored the book *Wanghong as Social Media Entertainment in China* (2021). He also published articles on Chinese creative industries, platform studies, and social media culture in leading academic journals. His research interests include cultural industries and creative labor, social media entertainment, platform studies, and Chinese contemporary culture. See also https://www.rug.nl/staff/j.lin/.

Miao Lu is a Postdoctoral Fellow in the School of Journalism and Communication at the Chinese University of Hong Kong (CUHK). She holds a PhD in Communication from CUHK (2020). Her research resides at the intersection of critical media studies, STS, and ICT4D. Her dissertation investigates the role of a Chinese phone company in Africa's digital transformation. Her work has appeared in academic journals such as *Media, Culture and Society* and *Chinese Journal of Communication*.

Daniel H. Mutibwa is an Assistant Professor of Creative Industries and Digital Culture in the Department of Cultural, Media and Visual Studies, University of Nottingham, UK. Daniel researches and teaches in the areas of media and communications, creative industries, digital culture, arts and citizenship, and transformations in communities and culture. He is the author of *Cultural Protest in Journalism, Documentary Films and the Arts: Between Protest and Professionalisation* and coeditor of *Communities, Archives and New Collaborative Practices* (2019 and 2020 respectively).

David Nieborg is Assistant Professor of Media Studies at the University of Toronto. He holds a PhD from the University of Amsterdam and held visiting and fellowship appointments with MIT, the Queensland University of Technology, and the Chinese University of Hong Kong. David has published on the game industry, apps and platform economics, and games journalism in academic outlets such as *New Media & Society*, *Social Media + Society*, and *Media, Culture and Society*. He is the coauthor of *Platforms and Cultural Production* (2021) with Thomas Poell and Brooke Erin Duffy. See also http://gamespace.nl.

Jack Linchuan Qiu is Professor at the Department of Communications and New Media, National University of Singapore. He has published numerous books in English and Chinese, including *Goodbye iSlave: A Manifesto for Digital Abolition* (2016), *World Factory in the Information Age* (2013), and *Working-Class Network Society* (2009). Jack is an elected Fellow of the International Communication Association (ICA), a recipient of the C. Edwin Baker Award for the Advancement of Scholarship on Media, Markets and Democracy, and the President of the Chinese Communication Association (CCA).

Serra Sezgin is Assistant Professor of New Media and Communication at Ankara Science University, where she has been serving as department head since 2020. She received her MSc in Communication Sciences from Hacettepe University (2013). She holds a PhD in Journalism from Ankara University (2019), where she had worked as Research and Teaching Assistant for six years. Her doctoral thesis has been published as an e-book, *Digital Games Ecosystem: Creative Industry and Labor* (2020). Her research interests focus on new media, creative industries, labor, and game studies.

Bingqing Xia is a Lecturer in the School of Communication, East China Normal University, Shanghai, China. Her research focuses on new media studies, particularly digital labor research. She has published several papers on digital labor issues, such as the conditions of work in China's internet industries. She is now working on a book about working life in China's internet industries, and conducting fieldwork in China's AI industry to explore data labeling labor's working life.